V-6 PERFORMANCE

V-6 PERFORMANCE

EDITED BY
LARRY SCHREIB

PRODUCTION BY
LARRY ATHERTON

ISBN 0-931472-13-X
PART No. 13-X

S-A DESIGN BOOKS, 515 WEST LAMBERT, BLDG E, BREA, CA 92621-3991

CONTENTS

FOREWORD

This book was written during a period of intense V-6 engine development and experimentation, both on the part of auto manufacturers and the specialty equipment aftermarket. The text had to be revised and updated literally as it was being written. New developments are taking place, new parts are being produced, and even entire new engines are being introduced (such as a 4.3-Liter V-6 diesel from GM) as we put this book to press. Consequently, we have had to speak in rather general terms about certain aspects of V-6 modification, plus we have discussed several developments that remain in the "proposed" or "scheduled" stage—such as certain factory or aftermarket performance parts—as of our deadline. To the best of our knowledge all such information is accurate and reliable, but we can never be totally sure of a part's availability or its final form until it is actually handed over the counter. Our intention in discussing such products is to demonstrate the rapidly evolving current developments in V-6 technology and to give you a good idea of what to expect in the near future. If you are in the middle of a V-6 project, however, be sure to ascertain that specific components are indeed available, and that they have not been changed from their description in this book, before you build the rest of the motor around them.

Secondly, wherever possible we have included manufacturers' part numbers to define products discussed here. Although we have taken great pains to make sure they are correct, such numbers are only as accurate as our sources, they may be changed or superceded after our publication date, and they have an uncanny knack for falling prey to typographical errors. When ordering a part described in this book, be *sure* to double check with the manufacturer or local dealer to insure that the part is still available and that the part number given accurately identifies the specific part you want.

Finally, we would like to thank several individuals who went out of their way to help in the compilation of this book and who gracefully withstood our persistent questions: Jim Bell of Kenne-Bell Enterprises, Joe Negri of Buick Motor Division, Doug Roe, Jim Ruggles, Dennis Novotny of TRW, Jerry Spotts, Richard Lyndhurst, Bill Smith of Yenko Performance, Bob Milliken of Hank the Crank, Shelley Arkow, Bill Jenkins, and Bob Mortensen of MacPherson Chevrolet.

THE AUTHOR

A book like *V-6 Performance* is produced by someone with uncommon abilities. To successfully collate and narrate the endless technical details included in the following pages, the author must be more than a facile and diverse writer, he must also be a tenacious researcher, an imaginative photographer and a knowledgeable car buff. Pat Ganahl accomplishes this—and much that is as difficult but less obvious—with the disarming flair of a California beachboy.

Pat's first car was a worn out '48 Chevy four-door sedan, into which he promptly dropped an early-style 216-inch Chevy straight-six. Today, nearly 20 years later, he still drives this machine daily. Over the years Pat has built ten engines—all Chevy inline sixes—for the venerable '48, and the current engine—an early model 235-incher with three carbs, Offenhauser manifold, McGurk cam, Fenton Headers and a Mallory ignition—has hauled home a few trophies from local antique drag races. Though it may not be considered an exceptional hot rod by the trendy standards of modern street-rodding, it is a very unique machine, because Pat has done everything—from engine assembly to the impeccable body and paint work—with his own hands, in his small backyard garage.

With this background (and a Master's degree in English Literature), it is not surprising that Pat's freeform talents have been widely acknowledged in the U.S. and abroad. For five years he was the editor of Street Rodder Magazine, and he is considered somewhat of an authority on early street-rodding history. His best-selling book *Ford Performance* (published by S-A Design Books) is an oft-quoted reference among Ford enthusiasts, and his familiar byline has appeared on many articles in Hot Rod Magazine, Car Craft Magazine, Popular Hot Rodding Magazine and various special-interest publications too numerous to list here. In addition, he produces a monthly street-rod column for several automotive magazines in England and Sweden. And sandwiched between all of this, was a three-year stint as the Southern California travel and outdoor-sports editor for Sunset Magazine, a prestigious journal of western lifestyle. (A position that expanded his prolific knowledge of So Cal surfing beaches.)

If there is such a thing as a quintessential street-rodder, Pat Ganahl qualifies. In an era of ever-increasing high-tech complexity, Pat's idea of performance is a throwback to the no-nonsense, do-it-yourself approach of bygone days. He feels that in today's world—as in the past—knowledge, ingenuity and craftsmanship are more important than spending money.

INTRODUCTION

The V-6 is the new kid on the block. Internal combustion engines have taken a variety of other forms throughout piston engine history. Straight fours, opposed fours, straight sixes, flat sixes, straight eights, V-8's, V-12's, and V-16's have all been standard configurations for automobiles. Though pioneered by Lancia and Ferrari in the 50's, then introduced briefly to Americans by Buick in 1961, it was not until the mid-1970's that the V-6—an apparently logical design—re-emerged suddenly and decisively. In the United States millions of the "bent sixes" have already been produced, with more rolling off the assembly lines every day. This book will examine four of these V-6's in detail, the 90-degree Buick (both even- and odd-fire), the 90-degree Chevrolet, the 60-degree Chevrolet (optional in all GM "X-body" front-wheel-drive vehicles), and the 60-degree Ford/Capri. Although not discussed at length here, V-6's have also been put into production recently by Alfa Romeo, Maserati (Citroen), and by a joint development venture by Peugeot, Volvo, and Renault. And for 1982, Ford introduced an all-new 90-degree V-6 that has been under development for several years.

The V-6 engine is definitely here, and here to stay. As V-8 production is being phased out by all major manufacturers, and existing units inevi-

Performance is a matter of perspective and relative size. The Alfa Romeo GTV-6 might be considered a "compact" here, but a true muscle car in Europe. With a 152 cubic inch (2.5-liter), 60-degree V-6, it puts out 125 horsepower and runs 0-to-60 in 8 seconds with a top speed of 125 mph.

tably become depleted, the V-6 will remain as the "big" machine, the muscle-motor if you will, of the future. Where has it come from, why have we got it, and what can we do with it in terms of modifications for performance and economy?

When Buick developed and introduced the 198 cubic-inch Fireball V-6 in 1961 they were both prescient and optimistic. But the time wasn't right. The minor economic depression that spawned a fleet of American small cars in 1960 passed quickly. American buyers rebounded typically, demanding the biggest and most powerful

engines ever produced and hungrily supporting our last great muscle-car era. Buick eagerly sold its V-6 engine and tooling to Jeep/AMC in 1967, while the V-6's parent aluminum Buick V-8 went to Rover of England.

But times change. The Arab oil embargo of 1972 underscored the writing on the wall. The days of cheap and plentiful gasoline will never return, nor will 400 horsepower production engines. Usually Detroit is slow to make major powertrain changes. But in '72 governmental regulations and skyrocketing gasoline prices escalated the transition. Muscle motors disappeared overnight. Compression ratios plummeted. Sales of 4-cylinder compacts soared.

By 1974 Ford was installing a 2800cc version of their European-designed 60-degree V-6 in Pintos and the downsized Mustang II. Ford had begun importing this engine from Germany in 1972 in the Mercury Capri. In 1975, Buick was more than happy to buy back rights and tooling for their 90-degree V-6 from Jeep. It went immediately into Buick Skyhawks and Oldsmobile Starfires. The next year, GM began using the V-6 in most of the Buick, Oldsmobile, and Pontiac small and midsize cars. Chevrolet also relied on the Buick bent motor until it could complete development of its own V-6, patterned after the proven smallblock Chevy V-8, for introduction in the 1978 Malibu (except in California). By 1980,

In Europe, where most cars are powered by four cylinders, a six is a big, exciting engine. A V-6 is a compact muscle motor that will squeeze into the place of a four. This Group-2 '74 Capri, driven by Jochen Mass and Niki Lauda, produced over 400 horsepower from its Cosworth four-cam, fuel-injected, 3.4-liter Ford V-6.

It has taken a while for Americans to discover the performance capabilities and driving excitement available from our smaller cars, but we will be seeing more race cars like this one in the future. The Yenko IMSA Citation, with a moderately modified 170-inch V-6, has 220 horsepower on tap in a light, nimble package.

Chevy had completed an all new 60-degree V-6 for use in GM's new X-body mid-size cars (Citation, Omega, Skylark, and Phoenix) incorporating front-wheel drive and transverse engine mounting. Although Ford ceased importing the German V-6 in 1981 in favor of the turbocharged version of their nearly-as-large OHC four, they should have a new American-built 90-degree V-6 in production by the time you read this.

WHY THE V-6?

The V-6 is in several ways an engine of compromises; most are positive, some are negative. But they add up to an engine that is right for the era of compromises in which we find ourselves. The factors that make the V-6 appealing for today's cars also make it the primary performance engine of the future, at least for as long as piston-type engines remain in production. We are ruling out, of course, some new-found wealth of petroleum deposits or a low-cost substitute for gasoline, both of which appear unlikely.

The surest way to reduce an automobile's fuel consumption is to reduce both its size and weight. This Detroit has finally done with our new "downsized" cars. Any good textbook on engine design will tell you that engines should be made to fit cars, not vice-versa. Today's small cars need small engines. Not just small in terms of cylinder displacement, but small in *external* displacement—in length, width, and height. So today, if we want an engine with more power, more cubic inches—for instance a six rather than a four—it has to fit into the same small engine compartment, and be nearly as light, as the four banger. A folded-up six fits. The V-6 is a surprisingly compact, attractive power package.

The second major consideration for production engine designers is cost of manufacturing. The V-6's are certainly not the first passenger car engines to suffer compromises in this regard. In the case of our current V-6's, there was also an urgency to get the engines into production as quickly as possible, which meant borrowing from existing designs and using existing tooling if possible. Consequently both Buick and Chevrolet patterned their V-6's after existing V-8's. This eliminated lengthy (and today very costly) research and development time necessary for an all-new engine, plus it meant the new engines could be built on the same machining fixtures already in place at their factories, and could use many interchangeable V-8 parts. The Ford V-6, snapped up ready-to-use from Ford of Germany, had been designed there in conjunction with a 60-degree V-4 engine introduced in the early '60's. Even in the case of the Peugeot/Volvo/Renault V-6, initial design was predicated on developing a companion V-8 from the same patterns (hence the 90-degree cylinder angle), though the energy situation eventually precluded the release of the V-8.

Why weren't V-6's built long ago? To begin with, early cars had plenty of engine room, and their shape was better suited to long, tall, and narrow power plants. Secondly, the straight six is an inherently balanced engine, while the V-6 is not quite perfect in this respect. Third, straight sixes are less expensive to build than V-motors (the I-6 has only one head and one bank of cylinders to machine). For these reasons, the major manufacturers developed straight sixes long ago, re-

Now that Chevrolet has put the 60-degree motor in the S-10, the V-6 should become a very popular swap package for other small pickups. The Corvette-style valve covers were custom made for a factory experimental model.

Slipping a compact V-6 into an early Ford street-rod chassis is simple. With typical street performance modifications, like a 4-barrel intake, headers and a mild cam, this 3.8-liter Buick will give the Model A plenty of power plus 25-30 mpg economy.

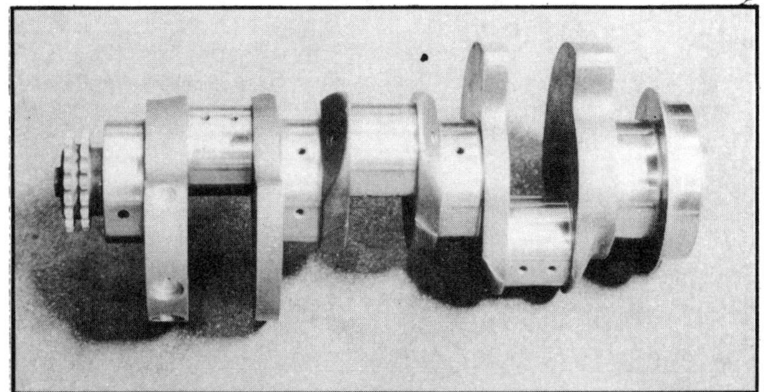

Methods for hot rodding a 90-degree V-6 differ little from familiar practices developed over the years for V-8s. The 4.1-liter Buick in this HOT ROD Magazine '32 Ford roadster was given the full treatment by Smokey Yunick. Look closely and you will see Smokey's modifications to the oiling and cooling systems.

Short and stout is an apt description for a V-6 crankshaft. This one is from a 90-degree Maserati, an odd-fire motor.

fined them over the years, and had little reason to change designs.

But Detroit has finally been forced into a period of changes. The public is now ready for a fleet of small, light, and *efficient* vehicles. They don't need V-8 power. Straight sixes won't fit under their hoods. Four cylinders are fine, especially for economy. But the light, solid, healthy V-6's will not only slip into the new small cars, they'll supply them with enough power to keep "performance" a part of our automotive vocabulary.

Would Detroit consider a radically downsized, small-displacement, light aluminum V-8 as a performance engine for today's cars? Such an engine would be slightly superior over comparable V-6's in terms of balance, smoothness of power pulses, and even volumetric efficiency. But designing and tooling such an all-new engine would be prohibitively expensive, plus it would cost more to produce on a per-unit basis, as compared to the V-6's we now have.

V-6 VERSUS STRAIGHT SIX

A recent ad for one of the major manufacturer's V-6-powered cars went something like this: "If you like the economy of a six, but want the power you associate with a Vee engine design, then our new V-6 is for you."

Does a Vee engine have more power than an inline? Certainly not. Cylinder angles have no relation to power produced. If engine size and weight didn't matter, most designers would choose the inline over the Vee configuration for a six because of the inline's inherent balance. Theoretically, the Vee design does not increase power.

But there are other differences that can affect net power or efficiency. Right off the top, if the V-6 saves 100 to 150 pounds over an equal displacement inline, the weight the V-6 doesn't have to pull translates to extra usable horsepower. If the V-6 is lower in overall height, the car can have a lower, more aerodynamic hoodline, which converts to extra usable power at highway speeds. And if the V-6 is up to eight inches shorter than the I-6, this is space that can be removed from the car's frame and body, thereby further reducing its weight. Naturally all of these factors improve fuel economy at the same time.

Structurally, the Vee design makes a much more rigid engine than the inline. The straight six block gives support in the vertical plane, but hardly any in the horizontal. Whereas the V-6 suffers some inertia vibrations, the I-6 has always been plagued with crankshaft torsional vibration and flexing. The extremely short length of the V-6 crank adds to its strength. Current I-6's use seven main bearings to reduce

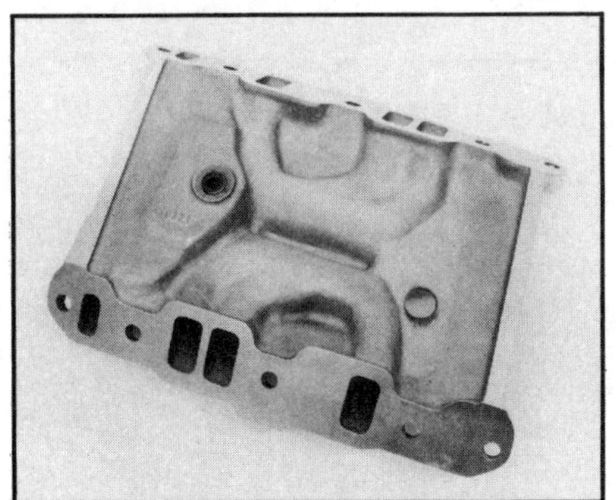

The underside of a V-6 intake manifold— this is an Offy 4-barrel for a Buick— clearly shows the "dual log" single-plane design.

crank bending, but this increases frictional losses; V-6's need only four mains.

The straight six also suffers twisting and bending of valvetrain components (cam, pushrods) because of their long length. Plus most inline-six valvetrain parts (especially pushrods and rocker-arms) are bigger and therefore heavier than V-6 or V-8 counterparts.

The inline six is a plumber's dream for fitting tuned exhausts (the V-6 fails in this category), but the V-6 holds a decided advantage on the more critical intake side. Inline sixes have always had intake problems, especially with one carburetor. An equal length runner manifold is out of the question. And with multiple carbs, straight sixes always seem to run lean at the front cylinders, rich at the rear. Intake manifolding on a V-6, on the other hand, is one of the engine's real plusses. With alternate bank firing order (1, 6, 5, 4, 3, 2) the engine has evenly-spaced intake pulses in each head. Therefore it can use a simple log manifold, one half feeding the right head, the other half feeding the left head, rather than a dual-plane intake as used on most V-8's. A single 2-barrel (or 4-barrel) in the middle can feed the ports through nearly equal-length runners, each half of the carb serving half of the engine.

Beyond these points of contrast between Vee and inline sixes, one big reason why the V-6 will appeal to today's engine builders and tuners is that, in most ways, it goes together and responds to modification like the V-8's we are familiar with. In the past twenty-five years we have amassed a wealth of technical information and practical experience with our current V-8's, especially the smallblock Chevrolet. Much of this expertise will translate directly to the V-6 engines under discussion. Better yet, a significant number of tried and proven performance components will swap directly into the 90-degree sixes, or can be easily adapted to them. The V-6 is not only an engine that fits today's cars, it's an engine that fits our experience as well.

BALANCE AND VIBRATION IN V-6'S

Piston engines, which convert work generated in a linear motion to work delivered in a rotary motion, vibrate. As each piston goes up and down (that is, reciprocates), it tends to shake the engine. Think of a one cylinder engine and the effect is obvious. The engine will have a "shake," or "inertia force," acting up and down in line with the cylinder. If you add a second cylinder next to the first, so that one piston moves up as the other moves down, their inertia forces will offset each other, and the engine will no longer "shake" up and down as a whole.

However, since these two cylinders must be spaced some distance apart, their opposite inertia forces will also rock the engine front to back. Imagine you are holding a shaft in the center, and opposing forces push up at one end, down at the other, then reverse. This type of force—a pair of equal and opposite forces acting on a shaft at a distance from each other—is called a couple. This is another type of vibration, or unbalance, common to piston engines. The example given is called a rocking couple. If we add two more cylinders, and arrange the crank throws so that the two outer pistons move up while the middle pistons move down, the arrangement cancels out the rocking couples in each pair of cylinders, balancing the engine.

Also consider the rotating crankshaft. If it has one throw, it will be unbalanced as it spins. However, a counterweight can be attached to the crank, opposite the throw, to balance it. If the crankshaft in a two-cylinder engine weren't counterweighted, the centrifugal forces of the two arms would create a couple tending to rotate each end of the engine about a point in the middle of the crankshaft. Again, this rotating couple can be counterbalanced by a pair of bob weights on the crank. If it were not counterbalanced, the resultant rotating couple would combine with the rocking couple caused by the reciprocating parts to produce a more complex unbalance—probably a rotating couple of varying magnitude (that is, a rotating couple that would be greater in the vertical direction than in the horizontal). Now, if we add two more cylinders as above, with all crank throws spaced 180 degrees apart, the rotating crankshaft will balance itself without the need of counterweights.

These are the two kinds of motion which can cause vibrations in a piston engine: reciprocating and rotating. In multi-cylinder engines, either can act as a simple force or as a couple. Rotating unbalance forces, caused by the crankshaft and the big ends of the connecting rods, are simple and can be easily balanced. Reciprocating unbalance, caused by the acceleration and deceleration of the piston assemblies and the small ends of the rods, can be much more complex, especially in Vee engines. Either, or both, types of forces can be balanced in most common engine configurations either by crankshaft counterweighting or by arranging cylinders and crank throws so that forces cancel each other.

Engine balance would be relatively simple if that's all there were to it. However, since the piston is attached to the crank throw by a link—the connecting rod—the piston does not accelerate in simple harmonic motion like the crank throw does. The piston actually travels more rapidly through the upper half of its stroke than it does through the lower half. Consequently,

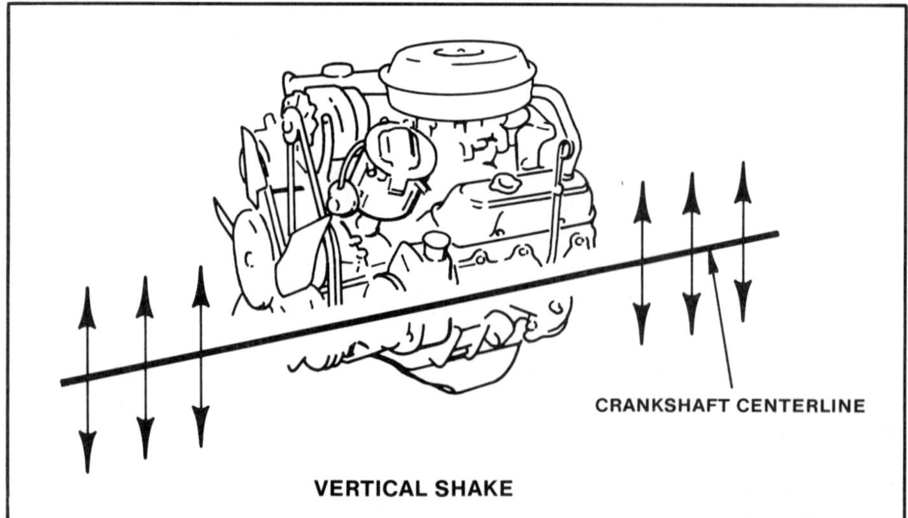

Unbalance forces caused by reciprocating or rotating components in a multi-cylinder engine can take many forms. A "shake" moves the entire engine, usually up and down (vertical shake), and is common in one-cylinder and four-cylinder inline engines: uncommon in Vee designs.

A "rocking couple" pivots the engine back and forth in one plane, usually horizontally or vertically.

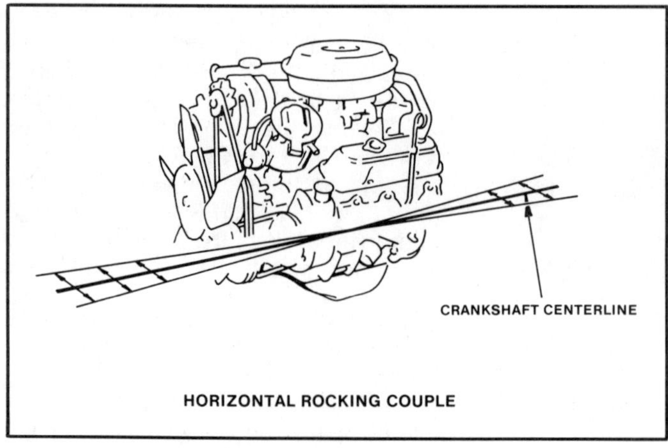

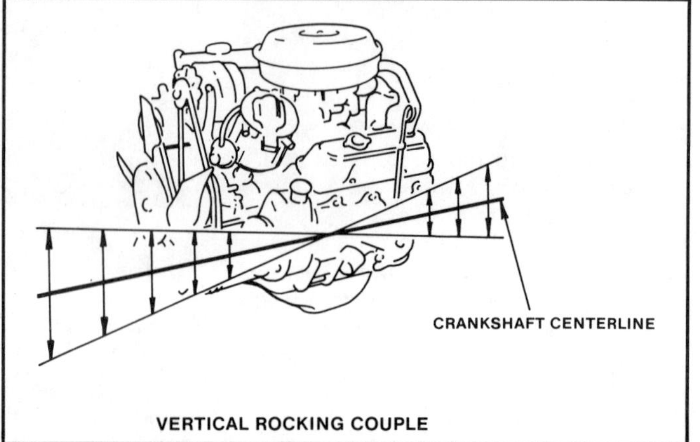

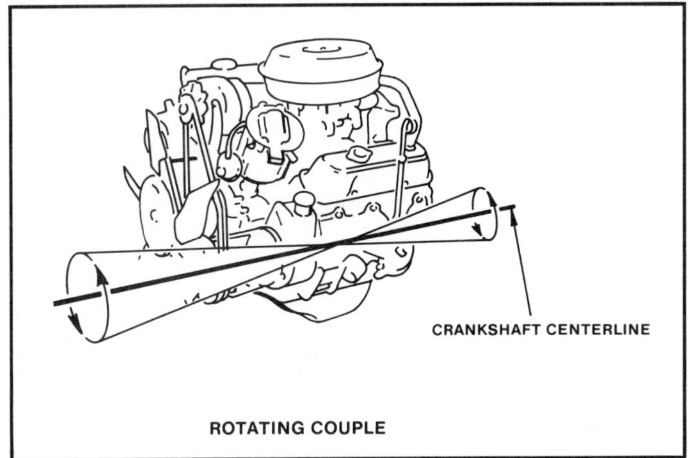

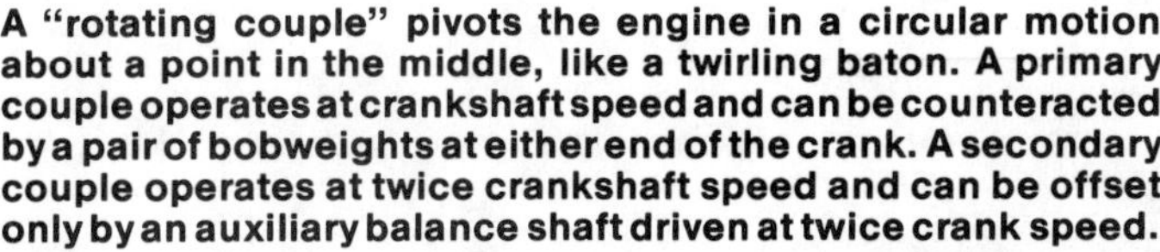

A "rotating couple" pivots the engine in a circular motion about a point in the middle, like a twirling baton. A primary couple operates at crankshaft speed and can be counteracted by a pair of bobweights at either end of the crank. A secondary couple operates at twice crankshaft speed and can be offset only by an auxiliary balance shaft driven at twice crank speed.

If an engine produces a vertical rocking couple of one magnitude and a horizontal rocking couple of a different magnitude, the result will be a varying rotating couple which rocks the engine more one way than another. Such an unbalance cannot be fully counteracted by any practical means.

analyzing reciprocating forces that act on an engine because of piston acceleration and deceleration becomes quite complicated. To figure these forces, engine designers use the formula:

$$F = \frac{W_c X^2 R}{g}(\cos A) + \frac{W_c X^2 R}{g}\left[\frac{\cos 2A}{\frac{L}{R}}\right]$$

where W_c equals the reciprocating weight (piston assembly plus reciprocating portion of the connecting rod), X equals crank angular velocity in radians per second, A equals the angle of crankshaft rotation from vertical, g equals the acceleration of gravity, L equals connecting rod length, and R equals crank throw radius.

In actual practice, this force is broken into two components, which are treated separately. The first term of the equation:

$$\frac{W_c X^2 R}{g}$$

is called the primary reciprocating force. Since it varies directly with crankshaft speed, it can usually be balanced by crankshaft weights or crank-and-cylinder arrangement. The second term:

$$\frac{W_c X^2 R}{g}\left[\frac{\cos 2A}{\frac{L}{R}}\right]$$

is called the secondary reciprocating force. It varies, or changes direction, twice per crankshaft revolution. Therefore, if it cannot be canceled out by

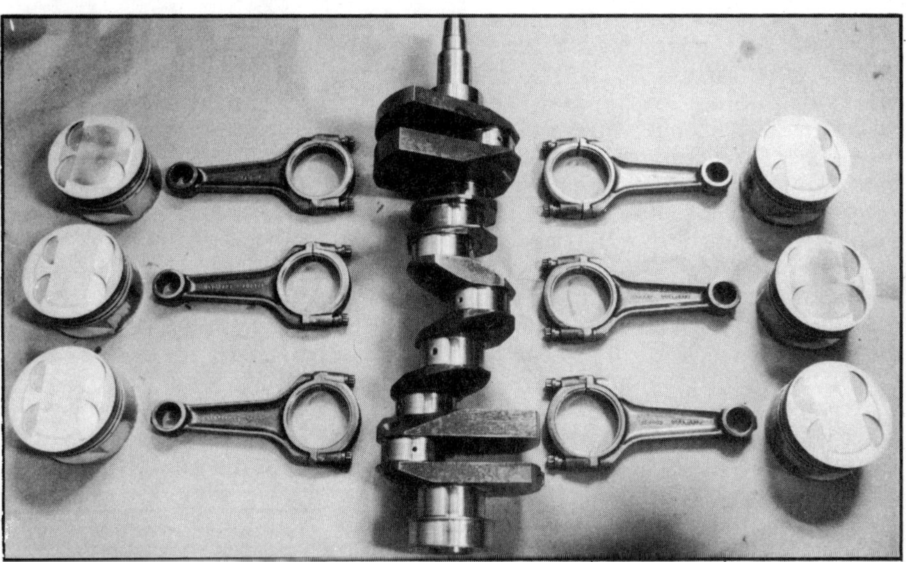

Large crankshaft bobweights—to offset a primary rotating couple—are obvious on this custom forged crank and reciprocating assembly made for the Cosworth Ford V-6 racing engine.

other cylinders, it cannot be balanced by weights on the crank. Secondary reciprocating unbalances, which can take the form of shaking forces, rocking couples, or rotating couples, can only be counteracted by one or two extra balance shafts (much like camshafts, with one lobe at each end) in the engine, geared to turn at twice crankshaft speed. Such shafts, and gearing to turn them, would obviously consume a portion of the engine's power. And their effect, though neutralizing the shaking or rocking of the engine as a whole, would not actually eliminate the unbalance forces on engine components (i.e., the crankshaft and bearings); they would only counteract them with opposite unbalance forces of their own. Fortunately, however, secondary reciprocating forces are only

about one-fourth the magnitude of primary reciprocating forces, and their effect on engine parts is usually negligible.

Obviously a full discussion of engine balance and vibration is far beyond the scope of this book, not to mention somewhat beyond the total comprehension of more than a dozen engineers, engine designers, and university professors we asked to explain it. (One noted Indy engine builder flatly stated, "Nobody understands it.") If you have a good working comprehension of physics and calculus, and you wish to pursue the subject further, there are a few references we would suggest. *The Internal Combustion Engine In Theory And Practice* by C. F. Taylor (M.I.T. Press, 1968) gives a very comprehensive and up-to-date analysis of the subject in Volume II, Chapter 8: "Engine Balance and Vibration." For a current, but slightly less complicated discussion, see *Engine Design* by J. G. Giles (Iliffe Books, London, 1968), pp 31-48. For a pioneering study of these forces as they affect V-6 engine designs, try to locate P. M. Heldt's articles in *Automotive Industries* magazine: "An Analysis of Vee-Type and Flat Six-Cylinder Engines" (Aug 15, 1948); "An Evaluation of the 90 Deg., V-Six Engine" and "Unique Balancing for 60-Deg. V-Six Engine" (both July 15, 1953). You should be able to find these works at a major public library or a university engineering library.

Understanding that these unbalance forces—rotating; primary reciprocating; secondary reciprocating... either as forces or as couples—can exist in various engine designs, let's examine possible V-6 configurations. To set the scene, let me first quote a portion of the first P. M. Heldt article mentioned above: "In the past, Vee-type engines have been built with two, four, eight, twelve, and sixteen cylinders [V-4's admittedly being quite rare]. If the six was passed by in the march of progress from Vee twins to Vee sixteens, there undoubtedly was a reason for it, and the most probable reason is that the Vee Six cannot be endowed with all of the qualities desirable in an automobile engine... . The objects usually aimed at when deciding on the number and arrangement of engine cylinders are uniform spacing of explosions and freedom from shaking forces and rocking couples. To these, it seems, we must now add a high degree of structural rigidity of the engine [necessitated by higher compression ratios; Heldt noted the strength of the V-6 as a major attribute over the I-6...engine size and weight were not as important in 1948]. Six is the minimum number of cylinders with which it is possible to obtain both uniform spacing of explosions and freedom from shaking forces and rocking couples without the use of special balancing devices. These qualities are inherent in the six cylinder in-line engine, and it is therefore only natural that as long as no special conditions have to be met, sixes always are built as in-line engines."

THE FLAT SIX

Like the inline six, the flat six or opposed six (180-degree V-6, if you will) is free of imbalance since opposing piston inertia forces act in the

The flat or "opposed" six is actually a 180-degree V-6. It is very short, but its width makes it impractical for use in front-engined cars. This is a rare Tucker engine, used in a rear-engined American car made in 1950.

same plane and counteract each other. Such an engine uses a six-throw crank with throws for each pair of opposing cylinders set 180 degrees apart. It's an excellent design—the crank needs no counterweights at all. However, the flat six is too wide to fit between the wheels and suspension of most front-engined cars. Consequently we have seen it used so far only in rear or mid-engined vehicles such as the Porsche and Corvair.

120 DEGREE V-6

Given the design parameters of evenly-spaced firing and/or minimal unbalance forces, that leaves three other workable configurations for V-6's. To get equal power pulses in a 4-cycle six, cylinders must fire at intervals of 120 crankshaft degrees. This criterium can be met in a 120-degree V-6 using a 3-throw crankshaft with 120 degrees between throws. This design allows two connecting rods to be attached to each throw, as in common V-8 engines, which is appealing both in economy of construction and reduction of engine length. However, the 120-degree V-6 is nearly as wide as a flat six, making it impractical for front-engined cars. Worse yet, the obtuse angle between cylinder banks (each bank actually

being a three-cylinder engine) places the inertia forces in different planes, where they cannot neutralize each other. The results in a 120-degree V-6 are both primary and secondary rotating couples of varying magnitudes. These cannot be fully counterbalanced by bobweights on the crank or by additional balance shafts in the engine. The only instances of 120-degree V-6's being used in automobiles so far have been the Type 156 Dino Ferrari of 1961, with which Phil Hill won the World Driving Championship, and the similar 1500cc, twin turbo, Formula 1 Ferrari introduced in 1981.

60 DEGREE V-6

The other design that allows even firing (with 60-degree spacing between crankpin pairs) is the 60-degree V-6. Such a configuration requires a six-throw crankshaft, which slightly increases both the length of the engine and the cost of production. However, although it sacrifices a bit in length, it is extremely narrow, the entire package being scarcely larger overall than a four-cylinder of equal bore and stroke. Given the confines of today's small cars, this design looks very appealing. Two of the engines discussed in this book are 60-degree

The 60-degree V-6 is extremely compact, being shorter and hardly wider than an inline four. However, the narrow distance between the heads causes some problems in designing a good intake manifold.

The 90-degree V-6 is a direct offshoot of the V-8 design, for which a 90-degree angle is "normal." It causes some balance problems, but offers excellent intake manifolding.

designs.

In terms of balance, the 60-degree V-6 produces both primary and secondary rotating couples due to reciprocating forces. But, since the primary rotating couple varies directly with crankshaft speed, it can be balanced by a pair of counterweights on the crank, one at each end. Since the secondary rotating couple varies at twice crankshaft speed, it could only be counteracted by a separate balance shaft turning at twice engine speed. (General Motors actually patented such a 60-degree V-6 in the early 1950's, but never produced it.) So the 60-degree V-6 is stuck with its secondary inertia unbalance. In a typical 60-degree V-6 operating at 4000 rpm, the magnitude of this rotating couple would be about 220 pound-feet. That's not much, and the effect is not very noticeable.

90-DEGREE V-6'S

The 90-degree V-6 with a simple three-throw crankshaft came about for two reasons. The first and probably most instrumental, as we have noted, was the pre-existence of designs, tooling, machinery, and parts for the 90-degree V-8's on which the V-6's could be patterned. Secondly, the 90° included angle between cylinder banks places the vector planes of the reciprocating forces at right angles to each other. Such an arrangement makes balancing of both primary reciprocating forces and couples possible with counterweights on the crank, opposite each throw, as in typical V-8's. However, the 90-degree V-6 also produces a secondary inertia unbalance, which takes the form of a horizontal rocking couple. That is, it tends to rock the engine back and forth horizontally, the front to the right, the rear to the left, then vice-versa, twice during each crank revolution. In a typical engine at 4000 rpm, this force has a magnitude of about 300 pound-feet.

But it is not this secondary imbalance that makes such a 90-degree V-6 vibrate so much. It is, rather, the uneven timing of power pulses that results from the crankshaft design. With three sets of crankpins spaced 120 degrees apart, and 90 degrees between cylinder banks, the pistons will not arrive at top dead center at regular intervals. With a firing order of 1, 6, 5, 4, 3, 2, as used in the odd-fire Buick, spacing between power pulses alternates at 150 and 90 crankshaft degrees (150°-90°-150°-90°-etc.). Obviously such an engine does not run as smoothly as an even-firing six. This unequal spacing creates a 1-1/2 order harmonic in torque output, which is not only disturbing to some drivers, but also creates torsional stresses in the crankshaft.

Buick apparently felt that engine smoothness was important enough to its customers to warrant an expensive remedy. To get the pistons to arrive at TDC at regular intervals, they "split" each throw on the crankshaft 30 degrees, thus converting a 3-throw crank to six throws without altering much else in the engine. The split-throw crankshaft, though unusual, is not unique; it was patented by Lancia in 1915. And, though this crankshaft is more expensive to manufacture, the solution was actually the most economical way out for Buick. Changing the cylinder block angle or lengthening the block and crank to a standard 6-throw configuration would have meant designing and tooling an entire new engine. The package they came up with, a 90-degree V-6 with even firing pulses, combines the optimum of compromises in terms of compactness, smoothness, and manufacturing economy.

There's only one problem—the new design upset the balance. Even

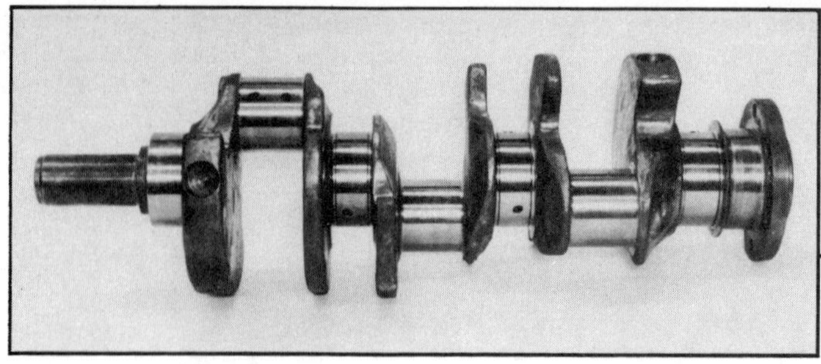

A comparison of odd-fire (top) and even-fire (bottom) crankshafts shows the novel solution developed by Buick for fitting six crank throws into the space of three.

V-6s are certainly not limited to cramped quarters. A modified "bent six," such as the blown Buick in this street-driven '55 Chevy, is light, powerful, economical, easy to work on and provides plenty of impact.

though opposing cylinders are still placed 90 degrees to each other, since their rods are no longer connected to the same crank journal, they no longer produce a neat, uni-directional, constant magnitude primary reciprocating force that can be balanced by a counterweight on the crank. The even-fire V-6 with 30-degree, split-rod journal pairs produces a primary rotating couple and a secondary rotating couple, both of varying magnitudes. (The secondary couple, for instance, varies from 183 pound-feet vertically to 318 pound-feet horizontally at 4000 rpm.)

Since the primary unbalance varies in magnitude as the crank rotates, only its average value can be eliminated by standard crankshaft balancing of 50% of reciprocating weight. Or, since the vertical and horizontal components of this force are related by inverse proportion, one can be increased or reduced (with the opposite effect on the other) by over-balancing or under-balancing the crank. Over-balancing can cancel the horizontal component, resulting in a vertical primary rocking couple of 470 lb-ft at 4000 rpm. Under-balancing (at 36.6%) cancels the vertical force, leaving a horizontal rocking couple of 470 lb-ft. Buick engineers chose the latter solution, deciding that the horizontal vibration would be the easiest to absorb with rubber motor mounts, thus producing the most pleasing overall package to passengers in the vehicle.

Chevrolet arrived at a different solution for their 90-degree V-6. Since the V-6 engine is, as we stated earlier, a melange of trade-offs to begin with, Chevy engineers decided to juggle several of these variables to come up with an interesting package. After plenty of drawing board and computer calculations, they decided on a practical, seat-of-the-pants test. Their theory was that a compromise between odd-firing and primary unbalance would produce the most pleasing package. So they built five different engines with crankshaft splay angles of 0, 15, 18, 22, and 30 degrees, installed them in vehicles, and gave them to design, development, and management personnel to drive. The democratic concensus was that the 18-degree engine was the smoothest overall.

Splitting the adjacent rod journals 18°, which gives firing intervals of 132°-108°-132°-108°-etc., reduces torque output fluctuations by 62% over the 3-throw "unsplit" crank (as in odd-fire Buicks). Then the engine was under-balanced (at approximately 46% of reciprocating weight) to reduce the vertical component of the primary reciprocating couple to one half its median value; this in turn increases the horizontal rocking couple a proportional amount.

All of these mechanical gyrations and manipulations probably sound quite confusing to the average engine builder or modifier. How much do you have to worry about it? Not much. The engines work, and they work well, even under harsh racing conditions—as has already been proven at Indianapolis and elsewhere. But it is important to be aware that various unbalance forces exist in V-6's, and that the Buick and Chevy 90-degree engines have been under-balanced to compensate for them.

The main point to consider is that the majority of manipulations undertaken by Detroit have been aimed primarily at passenger comfort. The average auto buyer wants a smooth-running engine. Minor engine vibration, and even torque fluctuation to a degree, are much less important to builders and drivers of performance machines than they are to the average auto consumer. Hot rodders and racers have long been familiar with, and willing to accept, all sorts of shakes, vibrations, and general lack of creature comforts as a trade-off for chassis stability and engine power. As one Grand Prix engine designer put it: "If the engine shakes, you just grit your teeth and drive through it. Of course, you try to design it so the vibration isn't in a critical rpm range." Devout Harley Davidson motorcycle riders wouldn't trade their knuckle-rattling, odd-fire V-twins for anything.

Do unbalance forces and vibrations of V-6's hurt their performance? Apparently not to any significant degree. Of course, any energy wasted on shaking the engine is energy that doesn't get to the driving wheels; but the amounts involved here are minimal. Of greater concern would be the effect on engine components—are the stresses going to wear or break engine parts? Again, it doesn't appear that they are of major concern. Unbalance forces will exert slightly greater pressure on main and rod bearings, which might make them wear a bit faster; but the effect does not appear significant in the millions of V-6's on the road. Even in high-rpm racing, where these forces can increase dramatically, no significant problems appear to have developed.

In comparing and contrasting the even-fire/odd-fire 90-degree V-6's of Buick and Chevy, the even-fire design appears preferable (though marginally so) for performance applications. Torsional stresses imparted to the crank by uneven torque pulses seem to be more detrimental than unbalance forces. Use of a heavier flywheel to dampen torque fluctuations (as is Detroit's practice) also reduces engine response. And the uneven firing sequence could cause some ignition problems.

Simply splitting crank journals to achieve even firing would have substantially reduced the strength of the crankshaft. Buick enlarged rod journal diameter to increase the crank cross-section in this critical area on the even-fire engine. Chevy retained the 2.10-inch pin diameter common to the parent V-8; this, and the .180-inch thick flange between journals, appears to give satisfactory strength and resistance to twist to the Chevy 18-degree split crankshaft. The extremely short

For Formula-1 Grand Prix racing Renault is using the same basic engine—with the addition of dual turbochargers and fuel injection—that powers Renault, Volvo and Peugeot passengers cars. This is a 90-degree, odd-fire motor.

length of the 90-degree V-6 crank is a major asset.

Two considerations for racers: the practice of offset grinding rod journals to smaller diameter to increase stroke length should be avoided; and since passenger comfort is not the main concern, these engines (90-degree Chevy and even-fire Buick) should probably be balanced at the standard 50% of reciprocating weight to reduce *overall* unbalance to its absolute minimum. For anybody changing things around on any of the V-6's, especially engine swappers, remember that small but constant engine vibrations, though not great enough to harm engine internals (or even be felt in the passenger compartment), can still fatigue and crack external parts such as mounts or brackets for accessories like alternators, air conditioner pumps, etc. Buick discovered that such mounts had to be strengthened on the original V-6.

V-SIX VERSUS I-FOUR

V-6's do not work out as perfectly in terms of balance as do straight sixes or V-8's. But these larger engines are fading from the picture. Buyers have stopped demanding them, and Detroit is cutting their production. So, rather than talking about the difference between V-6's and big engines, we should be comparing the V-6 to the

Europeans saw the performance potential in V-6 engines long ago. The 4-cam, 24-valve, Cosworth head conversion—designed ten years ago for the English Ford 60-degree passenger-car engine—was highly successful and actually beat Chevy V-8s for the Formula-5000 championship in England.

13

inline four. This is the engine it will be competing with in the future.

The V-6 looks significantly better on all counts. In weight, it's close. In power-to-weight ratio it's far superior. In overall size, they are nearly identical. And in terms of vibration, six power pulses—even or not—are smoother than four. Furthermore the inline four, a design the public has accepted since the Model T, suffers from a secondary vertical shaking force which cannot be balanced out without the use of a pair of rotating counterbalance shafts. Note that the vertical component is the one that Detroit designers feel is most upsetting to passengers. Yet the four-banger has always lived with this shake. Not until recently has any manufacturer felt it necessary to counteract the problem (Mitsubishi is now making a "balance shaft" four, used in Chrysler compacts). Frankly, today's engineers are probably going overboard trying to eliminate vibrations from the smaller engines. They may be overly sensitive to a public who has been spoiled on smooth V-8's. The subject is of engineering significance, to be sure, and any practical measures to minimize vibrations should be taken. But the emphasis given to the V-6's unbalance characteristics in previous technical papers—and even in this book—tend to make the problems sound much graver than they are.

THE V-6 IN THE SMALL CAR ERA

Since we are now caught, willing or not, in a period of automotive transition, let's look at the V-6 in a new perspective. Most of us are not yet weaned from the muscle car age. We've driven thumping, 400-plus-inch V-8's and we know what it feels like to stomp the throttle on one of those monsters. But gas was cheap and octane was high. Too many auto enthusiasts see the period ahead as an era of limitations, a downsizing of thrills as well as of cubic inches. But that's only one perspective.

Let's be a little more objective about those glorious muscle machines. They may have had 400 "advertised" horsepower, but it was encased in a big, lumbering, 3500-4500 pound vehicle. Contrast this to a modern downsized automobile as a *package*. You may have around 150 *net* horsepower in a stock V-6, but total vehicle weight has been trimmed by a thousand pounds in some cases. Not only is the resulting power-to-weight ratio close to the muscle car's, but the lighter car needs less muscle to get it moving from a dead stop (where fractions of a second count the most). Plus, it's nimble on twisting roads—a completely new automotive thrill for drag strip diehards.

For a case in point, compare the 1966, 427-inch, 4-speed Impala to the 1981 Chevy Citation X-11 with a 2.8-liter (170ci) V-6 rated at 135hp. Both cars will run the quarter mile in nearly the same time in street trim, yet the Citation will get better than double the fuel economy while emitting only a fraction of the pollutants. Now consider a race between these two vehicles through the esses. What it boils down to is that our performance cars of the present and future are much more efficient *packages*. Though we've pared down the muscle, we have also trimmed off the fat. The result is a new breed of street athletes—they're wiry and agile, they love to run, and they can keep it up on a trim diet.

Furthermore, during the muscle car era when Detroit was building our hot rods for us, the big-inch motors came fit to kill. They had rump-rump cams, big-port heads, hi-rise intakes, multiple carbs, high-compression pistons—hardly any further modification was necessary. Conversely, further modifying one of these engines netted little return. There wasn't much else you *could* do to them. The new V-6's, however, will respond dramatically to just a few of the basic hop-up techniques. You don't have to be a genius or a master mechanic with a full machine shop to increase the power of one of these little engines by 50% or even a 100%. Doing just that is what the rest of this book is about.

We should also state here that, at the time of this book's writing, ground is still being broken on V-6 engine development here in the United States. Two of the engines discussed here, the 60-degree Ford and 60-degree Chevy (the most unfamiliar design to American builders), are virtually untested in high-performance racing applications. A great deal of interest has been focused on the Buick, and more recently on the 90-degree Chevy, due in part to welcomed factory support for racing programs. The new "stock block" regulation supposedly to be imposed at Indianapolis (which limits turbocharged engines to 209 cubic inches), constant rumors of downsizing NASCAR sedans to V-6's, plus V-6 categories in both IMSA and SCCA sports car racing, have generated a recent flurry of interest in these engines. Detroit has responded with a rash of "heavy-duty" or "off road" parts, many of which are shown here. But some of these components are being outdated as quickly as they are

V-6s have been popular with street-rod builders for years because of their ease of installation, ample-yet-economical performance and undeniable good looks.

When shoehorned into a mini car—such as this Buick in a little Anglia—the V-6 sets the stage for a brand new form of performance street machine. A moderate-size motor in a tiny car will go just as fast as a big motor in a big car...for a lot less money.

being introduced. Others are definitely tentative at this point, or may not be released to the public through dealers' parts programs (even though some have been widely publicized in motorsports magazines).

Likewise, several major engine builders have been experimenting with the Buick and Chevy 90-degree engines, and their efforts have received considerable press. For the most part, these highly modified engines have been built for high-speed endurance racing (such as Indianapolis or NASCAR). Some of the theories being applied remain just that—theoretical (especially concerning engine oiling, block and crank support, water-passage modification, and cylinder-head reworking). In some cases the experts' theories conflict. We won't know what really works the best until hundreds of these engines are actually run in a variety of competitive events (look how much we are still learning about the smallblock Chevy V-8 and Chrysler hemi designs after all these years).

The major point to remember is that most of the "super trick" experimental work being done on these engines has been aimed (so far) at high-speed endurance applications. Drag racing technology, at this point nearly virgin territory for V-6's, may prove to be very different. For this book we have gathered a wealth of V-6 information from a variety of reliable sources, and we will attempt to portray the state of the art as it exists today. But we will also assume that the vast majority of our readers are primarily interested in V-6 modifications that apply directly to the street or to recreational racing activities. We will touch on high-tech pieces like dry sumps, block girdles, external oil systems, billet cranks, roller cams and so forth, keeping in mind their limited applications. But the major concern of this book will be to present the latest and most reliable in bolt-on components, minor modifications, and parts interchangeability to help you realize the most in practical performance from your V-6 engine without sacrificing any of its thrifty ways or much of your hard-earned cash.

A smaller engine means a lighter race car, and lighter race cars can run faster on fewer cubic inches. Although Ray Baker's V-6 Buick Skylark is only a prototype, this may well be the look and shape of NASCAR sedans of the future.

V-6's FROM EUROPE

The aluminum "sideways hemi" V-6 in this 1963 Lancia Flaminia uses a single, 3-throat Weber carburetor for induction.

No one seems to know exactly when or where the first automotive V-6 was built, although scattered examples turned up in a few motorcars from 1905 through the 1920's. Credit for the first modern V-6 engines goes to the Italians, particularly Vittorio Jano, designer of some of Italy's most famous racing engines.

LANCIA

It is not surprising that Lancia, a company which introduced a V-12 in 1919 and a V-8 in 1922, followed by thirty years of V-4's, would be a pioneer in V-6 design. In 1950 they unveiled

the Aurelia series, a car designed by Gianni Lancia and Vittorio Jano. It was powered by a 60-degree V-6 originally of 1754cc with a single camshaft in the block and pushrod-actuated valves. Combustion chambers were of a hemispherical design, with valves angled to the front and rear of the head (rather than side to side, as in a Chrysler) using L-shaped rockerarms. Most of the engine was aluminum. It was used in production sedans and in GT racing through 1959 in Aurelias, and into the 60's in Flaminias. A sports/racing version of this V-6 was also designed by Jano for Lancia in 1953. Designated the D23 and D24, it was raced only for a couple of years in limited numbers. This 3-liter engine, a 60-degree Vee, used two overhead cams per cylinder bank and two spark plugs per cylinder. A few were fitted with roots-type superchargers.

FERRARI

In 1956 Lancia could no longer afford "works" racing and they handed over all of their competition material, including designer Jano, to Ferrari. Since Ferrari had been using 60-degree V-12 racing engines since 1947, an engine of half this size would seem a natural development. Supposedly the V-6 design was suggested to Enzo Ferrari by his son Alfredo ("Dino"), who had begun work on such an engine before his untimely death. Ferrari named all subsequent V-6 models Dinos in honor of his late son.

The first Ferrari V-6 was introduced in 1956. Designed by Jano, it was a 65-degree Vee using dual overhead cams and dual spark plugs per cylinder. Called the type 246, this engine was built in several sizes (usually between 2 and 2.5 liters) and used in Formula 1 and 2 cars from 1958 through the 60's. In a similar design, but with a 60-degree Vee, it appeared in the Dino 246 GT street coupes of '68-72.

For 1961, after Ferrari switched to mid-engined cars, the type 246 engine was widened to a 120-degree Vee and called the type 156. Displacing 1.5 liters, it produced 190 horsepower using two 3-barrel Weber carburetors. With this engine in a Formula 1 car, Phil Hill won the 1961 Drivers' Championship and Ferrari took the Manufacturers' title. Ferrari re-introduced a new 120-degree V-6 for the 1981 F1 season in a dual turbocharged version.

In 1962 Ferrari introduced another V-6 with 60 degrees between cylinder banks and a single overhead camshaft per side, similar to their typical V-12's but cut in half. For induction, it used three 2-barrel downdraft Webers. A 2-liter version was called the type 196 SP, while a 2.8-L version was called the 286 SP. These were used only in limited sports racing and hillclimbs for a couple of years.

The 4-cam, 60-degree, transverse-mounted V-6 that powers mid-engined Dino Ferrari GT coupes is nothing less than a race engine in street clothes.

In 1981 Ferrari resurrected the 120-degree V-6 racing engine—in a dual turbocharged form—for Formula-1 Grand Prix cars.

The 90-degree, odd-fire Maserati V-6 used in front-wheel-drive SM Citroens uses four overhead cams and three Webers to produce 220 street horsepower from 183 cubic inches.

CITROEN AND MASERATI

Maserati had been building 90-degree V-8 engines since the 1930's. When Citroen of France commissioned them to build a V-6 for their new "Gran Turismo" Citroen SM luxury sports sedan in 1971, they came up with a 90-degree dual overhead cam six displacing three liters. Using a steel 3-throw crank, it had 90°-150° firing like the odd-fire Buick. For ignition, it used a single distributor with two separate sets of points, two rotors, and two coils—essentially two distributors in one. Three 2-barrel Webers were used for induction. The Citroen SM was built between 1971 and 1974, being imported to the U.S. only in '72 and '73. In 1975 Peugeot took over Citroen.

If you are looking for an exotic V-6 to use for an engine swap, the Renault-Volvo-Peugeot, all-aluminum, 90-degree motor would be an interesting choice, since quite a few have been imported to the US. It features hemispherical heads with single overhead cams and a cast aluminum oil pan which acts as a block girdle.

The new Alfa Romeo GTV-6 powerplant is an all-aluminum, 2.5-liter, 60-degree engine with single, belt-driven, overhead cams and hemi heads. Like most of the European V-6s, it produces over one horsepower per cubic inch in stock street form, yet delivers 28mpg on the highway in the GTV-6 coupe.

Maserati introduced the same engine in their Merak sports coupe for 1975. Supposedly an "economy" version of their Bora V-8, the Merak is still being produced in very limited quantities. The healthy 3-liter V-6 (183ci) produces a potent 220 horsepower at 6500rpm.

PEUGEOT/RENAULT/VOLVO

In a surprising joint venture by one Swedish and two French companies, a new all-aluminum V-6 of 2.6 liters was developed in 1975. Supposedly a similar V-8 was originally part of the plan, which may account for the 90-degree angle between cylinder banks, but the larger engine was scrapped when the trend towards smaller cars became obvious. The V-6 engine was designed for production passenger car use, and has been imported in large numbers to the U.S. in Volvo and Peugeot sedans. It has also been used recently in a turbocharged, dual overhead cam version by Renault for Grand Prix racing. Plus this is the same engine now being installed in the all-new, stainless steel, gull-wing Delorean being manufactured in Northern Ireland.

This 90-degree engine also uses a 3-throw odd-fire crank (cast-iron). The die-cast aluminum block has pressed-in steel cylinder liners, and a sturdy aluminum oil pan doubles as a main-bearing cap girdle. Chain-driven single overhead cams actuate canted valves

The Renault GP racing version of this engine—shown here with injectors removed— uses belt-driven, dual overhead cams.

by rockerarms. In this country the Volvos come with Bosch electronically-timed fuel injection, while the Peugeot uses an odd dual Solex carb setup with a primary single throat and a secondary 2-barrel.

ALFA ROMEO

The latest European entry into the V-6 field is Alfa Romeo. Their GTV-6 (the initials stand for Gran Turismo Veloce) coupe is powered by an all-aluminum 2.5-liter 60-degree V-6. This

is an even-fire engine using a steel six-throw crank. Belt-driven single overhead camshafts operate valves in the hemi heads in a unique manner: the cam is set above the intakes to run them directly, while short horizontal push rods and rockerarms actuate the exhausts. In Europe the engine is fitted with three 2-barrel downdraft Webers and produces 160 horses. The American version gets Bosch fuel injection and a triple catalytic converter for 154hp at 5500rpm.

2
CHEVY 60-DEGREE V-6

THE CHEVROLET 60-DEGREE V-6

In the introduction to this book we stated that the rationale underlying the concept of "V-6 Performance" is to fit a downsized powerplant into a proportionately downsized vehicle, then tune the package so that it will perform as well as, if not better than, the big V-8 cars of the past. Such was not only the thinking, but the stated intent of the Chevrolet Motor Division when they designed the 2.8 liter (173 cubic inch) 60-degree V-6 as a "power" engine option for their 4-cylinder Citation, as well as for the Oldsmobile Omega, Buick Skylark, and Pontiac Phoenix. According to the SAE Technical Paper published on this all new V-6 in July of 1979, *The General Motors 2.8 Liter 60° V-6 Engine Designed by Chevrolet* by David A. Martens, "...the performance goal was to at least equal that of a 1979 Nova equipped with a 5-liter V-8." Considering that the '79 Nova was itself a downsized vehicle to begin with, that's saying quite a bit.

The engineers obviously succeeded. To cite but one favorable "review" of the new V-6 Citation after a prolonged drive test (Hot Rod Magazine, August 1979): "with more than 6400 real world miles on its clock, the jury is back and the verdict is in. Yes, the

The new Citation X-11 with the 135hp HO-660 version of the 171-inch 60-degree V-6 is nothing less than a brand new type—that is, a new size and weight—of factory muscle car.

The new Chevy S-10 pickup with the V-6 option, a 4-speed and rear-wheel drive looks and acts even more like a factory hot rod. Jerry McDonald's offroad racer, packing a mildly modified 60-degree motor sleeved down to 146 cubic inches, has 220 horsepower on tap in a very reliable package.

Chevrolet Citation X-car is just as good as those early technical reports indicated. General Motors has indeed produced a true world class small car, designed and constructed in the only proper manner—from the ground up... We've been averaging over 22 miles per gallon...but the good news doesn't quit with economy. ...Our X-11's quarter mile times of 16.73 seconds and 80.08 mph, even under the load of every power-robbing accessory in the order book, compare quite favorably with most of the smallblock V8 machines currently produced in this country. Compared to imported compacts of its size and price range, the six-cylinder X-11 is practically a modern-day musclecar." Our point exactly, except that we would drop the qualifier "practically." And remember, this testing was with the early 115hp X-11, not the hopped up 135hp '81 model.

Of the four V-6 engines discussed in this book, the 60-degree Chevy is the newest, not only in terms of release date (it was introduced in model year 1980), but more dramatically in terms of design. Whereas the Ford was designed in 1961, and both the Buick and Chevrolet 90-degree engines were patterned on V-8 designs even older, the 60-degree Chevy is a completely new engine, one of the first really new Vee designs to come from General Motors in nearly two decades.

We stated in the introduction that the V-6 is a design that will allow a larger engine to slip into a space intended for a four-cylinder. The 60-degree Chevy is a fine example of this. It was made specifically to fit the front-wheel-drive X-body GM line which had been previously designed for an inline four—and the 60-degree included cylinder angle configuration was chosen to fit the larger engine into the narrow confines of the body.

But the more important point is that this engine was designed as a *power* option for these cars. The expressed

task of Chevrolet engineers, given the limiting parameters of government-mandated fuel economy and emission standards, was to squeeze as much usable power as possible from the little package. Using the latest design and casting techniques, every ounce of extra material was trimmed from the engine to save weight, but at the same time extra attention was paid to critical stress areas in order to achieve maximum strength. For example, rolled fillets on all crank journals increase the fatigue strength of this one part 100%.

Another example of the performance engineering devoted to this engine is the cylinder head design. The original test engine produced 108 horsepower. But the engineers went back to the flow bench, spending an unprecedented amount of time redesigning the intake port runners. The final port shape was actually *reduced* in size by 17%, which produced a 17% increase in airflow at maximum valve

The Yenko IMSA Citation, using high-compression pistons, cam, headers and a 2-barrel carb, has 230 horses on tap in the agile front-wheel drive body. Note X-11 cowl induction used on the race car.

lift. The production version of the engine nets 115 horsepower.

In the old days, for a 6% horsepower increase factory designers would have simply bored for more inches, bumped up the compression, or bolted on another carburetor. But the little RPO (Regular Production Option) LE2 V-6 is a perfect example of the new direction of automotive performance, and the ways in which Detroit engineers will work to achieve it.

After far too many seasons of horsepower cutbacks accompanied by stick-on decal cover-ups, Chevrolet has finally demonstrated that they still understand "blue bow tie" performance. If there were any doubts, the introduction of the H.O. 660 package for the 1981 X-11 Citation should put

them to rest. Without increasing the engine's size, they pumped a considerable 17.4% more net horsepower into the V-6. To do it, they did rely on a few of the old hot rod tricks—but they were carefully selected to increase power through efficiency. The package consists of higher compression pistons (8.9:1), bigger valves, a slightly longer duration camshaft, cool-air cowl induction, and a low-restriction exhaust system. These are some pretty solid factory muscle parts to back up the exterior decals and suspension handling package, and as we will see later in this chapter, they will probably remain—for the short run—a primary source of hop-up parts for non-H.O. 660 motors.

That's the silver lining; now we'll throw a couple of clouds into the picture. The first one affects all the GM engines discussed in this book, but the 60-degree more specifically than the others because its late arrival introduced it directly into the genesis of the automotive computer age. We're talking about the General Motors (more accurately Delco Electronics) Computer Command Control (CCC) engine monitoring and regulating system. Most California-bound 2.8-liter V-6's were fitted with this system in 1980, and all gasoline-powered GM cars and small trucks went on the computer in 1981. This of course includes the majority of all 60-degree V-6's produced.

The computer system, guided by a "brain" called the Electronic Control Module (ECM), reads the oxygen content in exhaust gases with electronic sensors, relays this data to the ECM, which in turn varies the carburetor mixture and the ignition timing. The specific purpose of the CCC system is to maintain a "perfect" stoichiometric air-fuel ratio of 14.7:1, primarily to keep engine emissions within government clean air standards. Besides varying the air-fuel ratio in the carb-

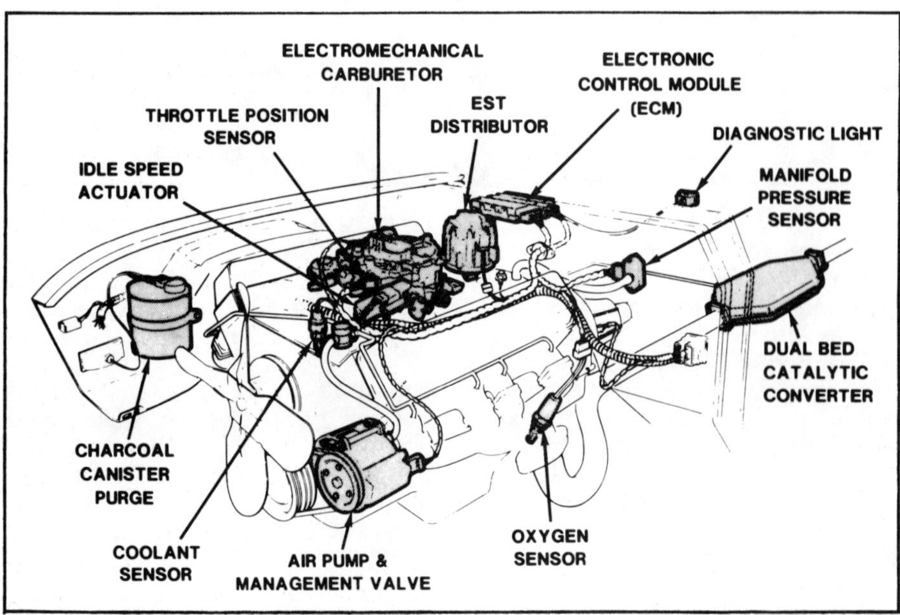

Although this diagram is for a V-8 car, the Computer Control System used on all '81-later GM passenger cars and light trucks are similar—and pervasive.

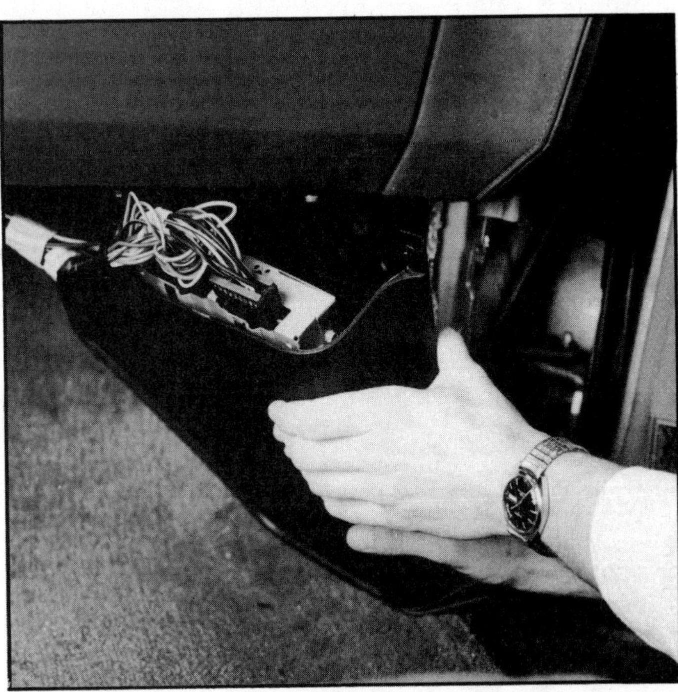

The Electronic Control Module (ECM) is a microcomputer that reads engine functions and tells it how to operate. As you can see, it's not something the typical hot rodder is capable of modifying.

The ECM microcomputer is mounted either in the right-front kick panel or under the dash behind the glove box.

uretor (or throttle-body injector) and the advance curve in the distributor, the ECM is also programmed to control idle speed and mixture, fresh air intake, the "air management system" (smog pump), and the torque converter clutch on certain automatic transmission models. As of 1981, the chip or PROM (Programmable Read-Only Memory) in the ECM could be programmed for some 2000 different functions. A specific PROM, with a set program, is installed in the ECM to match the body/chassis/engine/options package which it will govern. To cover the entire GM line the basic ECM brain can now be fitted with one of 200 different PROMs. The system seems to work quite well, perhaps better than the old "hand tuning" method.

The problem is that this new computer control system appears to be completely incompatible with our current bolt-on approach to street performance modification. *On an ECM-equipped vehicle, you cannot bolt on a set of headers, install a different camshaft, swap manifolds, carburetors, or distributors, or even add dual mufflers without disturbing the programmed balance of the "system."* Most any performance modification or parts swap will create some sort of change in both carburetor signalling and exhaust mixture. For example, a performance intake manifold will likely

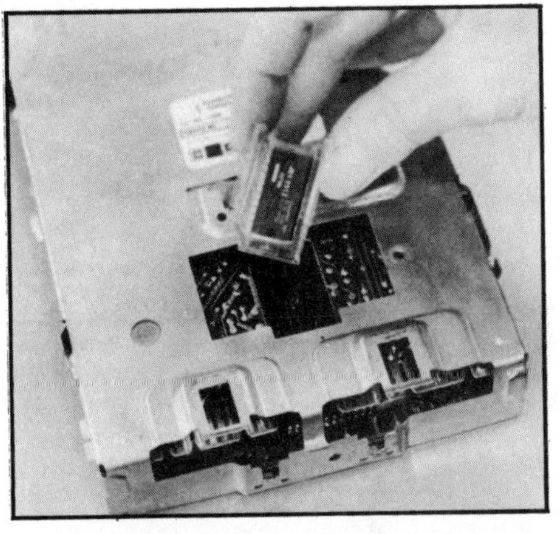

Although modifying the ECM is nearly impossible, changing its program chip (or PROM) is simple—just remove the cover, pull one out, and slip another one in. Nonetheless, you are limited to PROMs available from Delco—such as using an X-11 PROM in a car whose V-6 is modified to HO-660 configuration.

produce a stronger carburetor signal by increasing air velocity through the venturi. A performance camshaft will probably richen exhaust gases, while headers or dual exhaust will tend to lean the mixture. In previous hot rod practice, we would simply rejet the carb (or change venturi size) and recurve the distributor to balance the other changes we have made. With an ECM operating the system, however, the computer would constantly at-

tempt to reset both the carb and ignition to stoichiometric conditions—and could end up only compounding problems. Furthermore, with the ECM system, it is impossible to manually set (or change) carburetor metering or the distributor advance curve, since these are controlled by the computer.

At the time of this writing, the ECM system is brand new and yet to be explored by either hot rodders or the aftermarket performance industry. There may be a possibility that certain bolt-on performance parts can be developed to work compatibly with the computer. Or there may be a way to make the computer work compatibly with aftermarket parts (for instance by relocating sensors or mechanically modifying "robot" components). A longer shot, since it would require access to Delco's complicated program for the PROM (which they aren't about to make public), would be for the aftermarket to design "hop-up" chips which could easily be inserted in the ECM box. Such a PROM could be programmed to work with a specific cam, for instance, or with a specific combination of performance parts (i.e., cam, headers, intake manifold, and ignition). Or the PROM could be programmed to operate the factory components in a more "performance-oriented" mode. However, development of such aftermarket electronics—though it has been discussed—would be exceedingly difficult without Delco's direct assistance (an unlikely prospect), and will probably remain in the realm of the futuristic for quite some time.

That leaves the typical street performance modifier with two alternatives if his vehicle is computer controlled: leave the factory system the way it is and enjoy it, or else dismantle the *entire* system and ash can it in order to make any performance modifications at all. Since the latter alternative is the only one which allows performance modification to the engine—and that's what this book is about—we will gear our discussion to those enthusiasts who are serious enough about modifying their engines

and perhaps revolutionary ways to deal with this motor. The fact that these vehicles use the 60-degree V-6 in a "conventional" front-engine/rear drive configuration also means that this engine is now very adaptable to other body styles and will likely become popular for engine swaps, especially into other mini-trucks, small cars, and even early Ford hot rods. As the popularity of such 60-degree swapping grows, so should the market for bolt-on engine goodies (especially headers), since most of these engines will likely be transplanted without the ECM computer system.

In the meantime, however, there is just very little to discuss in the way of bolt-on parts for the 60-degree Chevy engine. We will tell you what we have

most of Katech's engine work on this V-6 in the following section. While performance products for the 60-degree motor remain quite scarce, any available from Chevrolet (such as H.O. 660 pieces, or heavy duty V-8 parts applicable to the V-6) as well as any emerging in the aftermarket, will be stocked by Yenko Performance and Marine, 575 W. Pike St., Canonsburg, Pennsylvania 15317 (phone 800-245-9909).

Another excellent source for 60-degree V-6 parts is MacPherson Chevrolet, 21 Auto Center Dr., Irvine, California 92714. They have just completed a two-wheel-drive S-10 pickup for SCORE Class 7 off road competition, to be driven by Jerry McDonald. Using an H.O. 660 V-6 block sleeved down to 2.4 liters (146 cubic inches) and modified only with a reground cam, ported heads, custom pistons, headers, ignition, and a 2-barrel carb, the engine produces a healthy 220 horsepower and proved extremely quick and reliable in its first long distance race. MacPherson's well-stocked and knowledgeable parts department specializes in Chevrolet heavy duty and "bow tie" performance pieces, and should have any available 60-degree V-6 performance parts in stock.

Although untested as of our deadline, a third "pioneer" 60-degree Chevy is being put together by midget racer Shelley Arkow of Canton, Ohio, and will also be featured in this section. Using direct-port fuel injection for methanol, plus high-compression pistons, a flat-tappet cam, and dry-sump oiling, Arkow expects to make 300 horsepower from the 173-inch motor for Outlaw racing (he will sleeve it down to 158 inches for legal USAC competition). However, the majority of the engine—block, crank, rods, H.O. 660 heads and HEI ignition—will remain basically stock. Arkow's expressed intent in switching to the V-6 is to build a competitve race motor for under $3000, using primarily production components. At this point the 60-degree Chevy looks like a potential winner in a low-cost, reliable, compact package that can compete with extensively-modified Volkswagens, expensive custom engines like the Sesco 4-cylinder, and production motors fitted with exotic custom cylinder heads. Only time will tell of its success.

BLOCK, CRANK, RODS

When designing the new RPO (Regular Production Option) V-6 Chevy engineers were primarily in-

Ohio midget racer Shelley Arkow and crew chief John Slomski are putting high hopes in the 60-degree Chevy V-6. Their goal in using the V-6 is to build a significantly less expensive competition engine using plenty of stock Chevy components.

to start by throwing away a $500 computer system that was included in the sticker price of the new car.

The second problem facing owners and builders of the 60-degree V-6 is compounded by the first. Virtually no aftermarket bolt-on performance pieces currently exist for this new engine—intake manifolds, headers, carbs, ignitions, and so on. And since the ECM system precludes the ready adaptability of such parts to these engines, the chances of seeing much in the way of new performance products for the 60-degree V-6 seem slim. On the other hand, the introduction of the 60-degree V-6 in the new S-10 Chevy compact pickup and the '82 Camaro means that the engine will be put in the hands of a broader, more performance-minded public. This should lead to a renewed effort on the part of the aftermarket to come up with new

learned from a couple of notable examples of seriously modified 60-degree motors, plus some factory experimental exercises, but what lies in the future for this little V-6 only time will tell. It has plenty of potential, and rodders are bound to find ways to tap it.

One of the pioneering efforts in Chevrolet V-6 modification is a front-wheel-drive Citation put together in 1980 for the IMSA Champion Spark Plug Challenge road racing series. Campaigned by Don Yenko, owner of a Chevrolet dealership in Pennsylvania long known for building, racing, and selling specially-prepared Chevies, this X-11 has proven very competitive in two years on the circuit. The 173-inch motor, assembled by Katech Engines (Mt. Clemens, Michigan), has undergone only moderate modification, yet delivers 230 horsepower with a single 2-barrel carb. We will detail

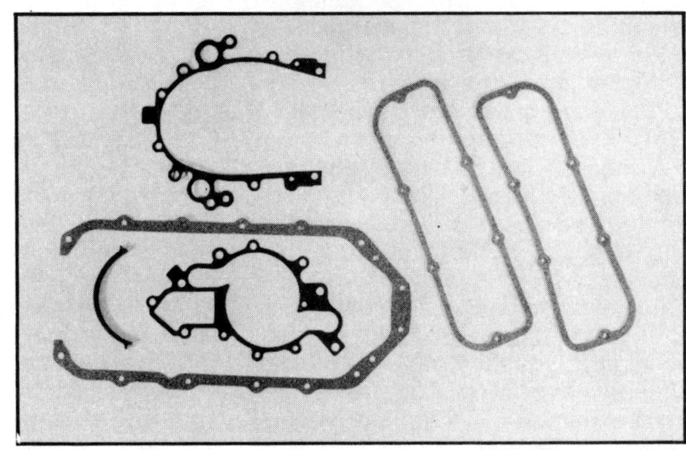

terested in a healthy powerplant in the smallest, lightest package possible. The overall bare engine (block and heads) measures roughly 15-inches long, 15 inches wide and 15 inches tall. Wherever possible, excess material was trimmed from the cylinder block—without creating any weak points. The bare block weighs in at just a hair under 100 pounds. Basic wall thickness is 4.5mm (.175-inch), and numerous lightening holes are cast in the top of the deck. Since the transverse engine in X-cars must partially support the transaxle, extra strengthening ribs and thickness were added at the rear of the block to produce a good rigid mounting flange.

Only two versions of the 60-degree block have been produced so far, both of 2.8-liter (171 cubic inch) displacement. The first is for transverse mounting in X-bodies; the second, introduced in 1982, is for rear-wheel-drive applications. Both blocks are essentially the same, but the latter version features some different bosses and bolt holes to accommodate both a

change in motor mount location and swapping the starter to the right side of the block. Both blocks have the same bellhousing bolt pattern, so the rear-drive transmission could be bolted onto an X-car block, but some modification (or fabrication of an adapter bracket) may be necessary to mount the starter on the other side.

We should mention that just about all of the 60-degree engine is built to metric dimensions, and uses metric fasteners. We will give equivalents in inches and pounds where applicable. Bore and stroke are listed as 89mm by 76mm, which works out to a neat 3.5 inches by 3 inches. Not incidentally, the 3-inch stroke is the same as Chevy's famous 265, 283, and early Z-28 302 V-8's, as is the connecting rod length of 5.70 inches, giving the 60-degree V-6 an identical rod length-to-stroke ratio as Chevy's performance-proven V-8 smallblocks. Nominal bore spacing in the 60-degree block is 112mm (4.41 inches). Crank center-line-to-deck measurement is 224mm (8.82 inches). Overboring possibilities

Although factory engines are assembled with RTV rather than gaskets, McCord has just released a complete standard gasket set for the 60-degree.

on this block are minimal due to thinwall casting. Chevy offers 1mm (.040-inch) as the largest overbore piston size; you might be able to go .060-inch or slightly more. The low placement of the cam in the block limits bore increases, even with new coring at the factory. Design parameters will allow 3mm (.120-inch) increases in both bore and stroke, resulting in a 3.2-liter engine—a good future possibility. But the game plan for modifying one of these mini motors does not by any means start with a quest for more cubic inches.

Although the majority of the 60-degree engine uses RTV sealant in place of gaskets (oil pan, valve covers, etc.), a composition-type head gasket is used on production engines. Chevy attempted using a steel shim head gasket at first, but found (despite improving the rigidity of both head and

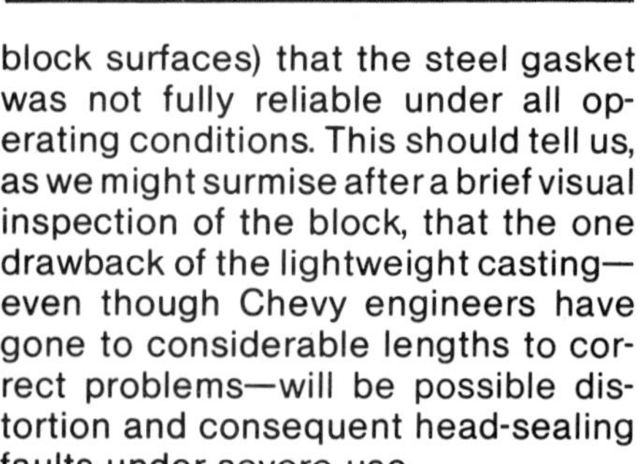

block surfaces) that the steel gasket was not fully reliable under all operating conditions. This should tell us, as we might surmise after a brief visual inspection of the block, that the one drawback of the lightweight casting—even though Chevy engineers have gone to considerable lengths to correct problems—will be possible distortion and consequent head-sealing faults under severe use.

The fact that the head attaches with only eight bolts (like the Ford 60-degree V-6) does not improve the situation. For performance use, especially with high static compression ratios or some form of supercharging, O-ringing of the heads would be a wise consideration. Sizing of the bores with a torque plate affixed (you may have to make your own) should go without saying on a serious build-up, and substitution of studs for head bolts is recommended (re-tap the block for SAE threads since metric studs will be hard to find).

The only other area where current builders have found a problem with the 60-degree block is the stock rope-type rear-main seal. Not only does it impart more drag on the crank than rubber-lip seals, but this one "goes away" at speeds over 6000rpm according to engine builder Byron Clemens of Katech. To solve the problem, he uses a Chevette 360-degree neoprene seal, which fits the V-6 crankshaft. To install this seal, however, the back of the V-6 block must be machined (counterbored) to accept it. Allow for .005- to .010-inch press-in fit. If you are building the engine for serious racing, you'll be doing other machine work on the block (such as align boring), and this extra step will be no big job at this stage. If you're using the V-6 on the street you'll very seldom be running at sustained high rpm; consider the installation of the Chevette rear seal—and the requisite engine tear down—

only if an oil leak becomes a chronic problem.

The fact that the 60-degree V-6 crank has a relatively small flywheel "shank," rather than a flange as used on most engines, allows the use of a 360-degree, one-piece oil seal. The small bolt circle for attaching the flywheel could conceivably lead to some problems, but none have surfaced so far. In all other respects the 60-degree nodular iron crankshaft appears to be an exceptionally stout and well-engineered piece. As we mentioned, all six individual crank pins and the center two main journals have deeply rolled fillets to increase strength. Rod journals are two inches in diameter, the same as early small-block V-8's. For serious competition, bearings should be upgraded to TRW/Clevite 77 type, which should be available from Yenko Performance, if not from your local parts store or Chevy dealer.

The Chevy 60-degree crank is bal-

anced with two large bobweights, one at each end, 180 degrees apart. To clear the oil-pump driveshaft, the rear bobweight was narrowed slightly and the difference is made up with balance weights added to the flywheel (or holes punched in the auto-trans flexplate). Consequently the crank and flywheel or flex plate should be rebalanced as a unit. So far racers have used standard balancing of the reciprocating assembly (at 50%) with no problems. The 60-degree design produces a secondary rotating couple unbalance, but the magnitude in this engine is 164 ft-lb at 4000 rpm—relatively small. However, if you are swapping this engine into another chassis, it would be wise to use stock-type motor mounts, which are designed to absorb this vibration.

If you are used to working on Chevy V-8's, be sure to note that this is the first Chevrolet V-motor to be designed with the right cylinder bank forward. Consequently the cylinders are numbered 1, 3, and 5 on the right side and 2, 4, and 6 on the left. This even-firing engine (120° crank rotation intervals) therefore has a straightforward 1, 2, 3, 4, 5, 6 firing order.

The stock connecting rods appear to be quite satisfactory for high-

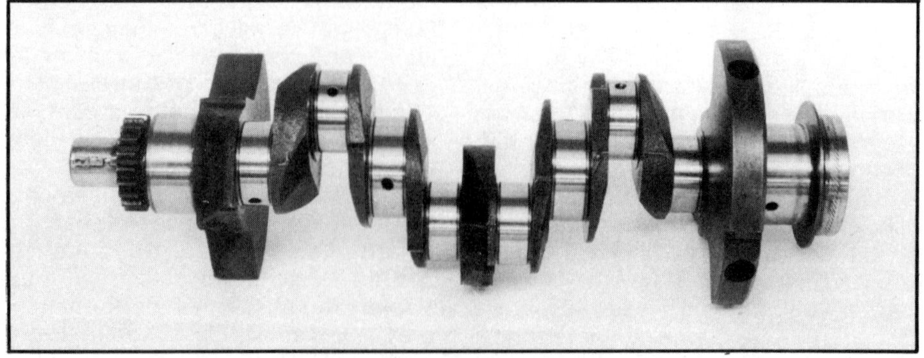

The stock, 6-throw, 60-degree crank appears extremely tough, especially since it features large-radius fillets on all main and rod journals. The only questionable part of this component is the short, small-diameter flywheel flange.

performance use. Other than fitting with upgraded bearings, as noted, and the usual prep of Magnafluxing, polishing, and shot peening, they should need no further work. It should be noted that the caps are held in place by 9mm metric bolts and nuts (approximately 3/8-inch). Since both the big end diameter (2 inches) and the center-to-center length (5.70 inches) of the V-6 rod matches that of the smallblock Chevy V-8, substitution of a wide variety of aftermarket V-8 rods into this engine should be fairly easy, though probably unnecessary except for specialized racing applications. What machining or clearance problems might exist with aftermarket V-8 rods remains to be seen, however, since very little information currently exists about such swaps.

The oiling system on this engine is another as-yet-unexplored area. The stock system is much like that in the 90-degree Chevy V-6: a rear-sump pan, the pump mounted at the rear main cap, a main oil gallery in the left bank intersecting the lifter bores, and annuli around the cam bearing bores to feed the main journals. No aftermarket oil system components are yet available, but the stock engine does

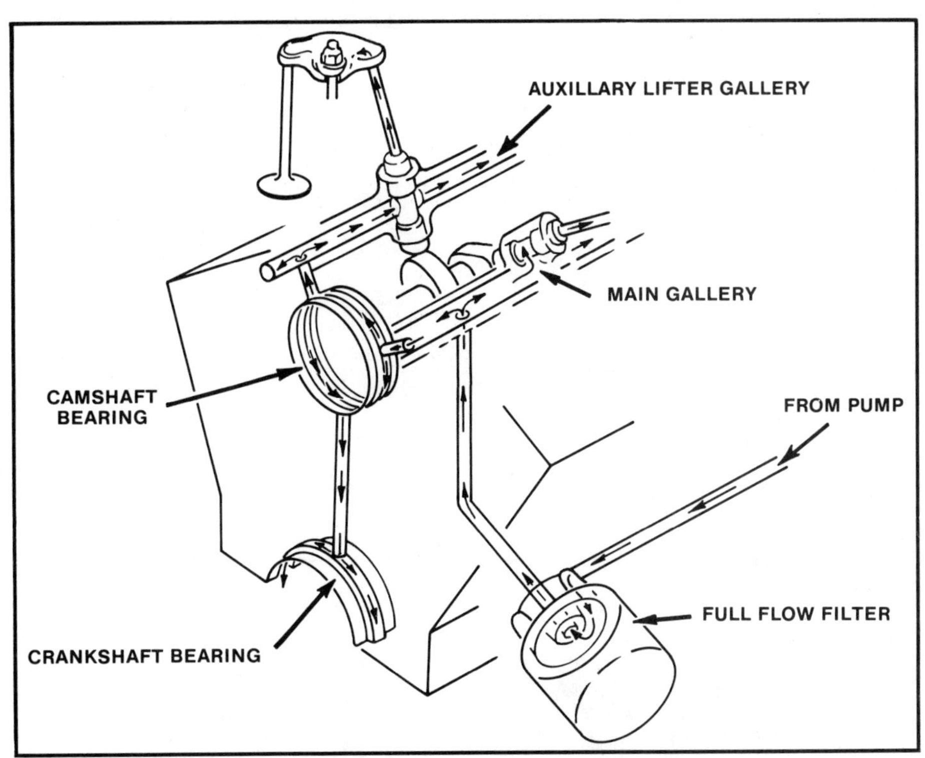

Other than the location of the filter, the oil system in the 60-degree is similar to that in the 90-degree V-6. The left lifter gallery is also the main oil gallery.

come with a windage tray as an integral part of the pan. The Yenko Citation uses an enlarged stock, wet-sump oil pan with trap-door baffles to control oil slosh on a winding road course, plus a bolt-on, V-8 windage tray modified to fit the V-6.

For circle-track racing some additional fabrication and testing may be required to insure adequate oil control. For example, Shelley Arkow's prototype V-6 midget uses a V-8-type dry-sump system with a custom-made Aviad pan and a typical external, belt-driven pump (with appropriate scavenge and pressure sections).

PISTONS

The stock RPO V-6 uses an aluminum, flat-top cast piston and the engine is rated at 8.5:1 compression. This piston has been used successfully in a few turbocharged or supercharged street applications with no problems, and the compression ratio is about right for the addition of a moderate boost (under 7psi) blower. However, water injection, ignition timing retard, or both under boost are probably necessary to save these stock pistons from detonation damage on available pump gasoline.

The 1981 H.O. 660 X-11 Citation comes with different pistons (part 14033129) which raise compression to 8.9:1. These should be available at your Chevy dealer (for about $30

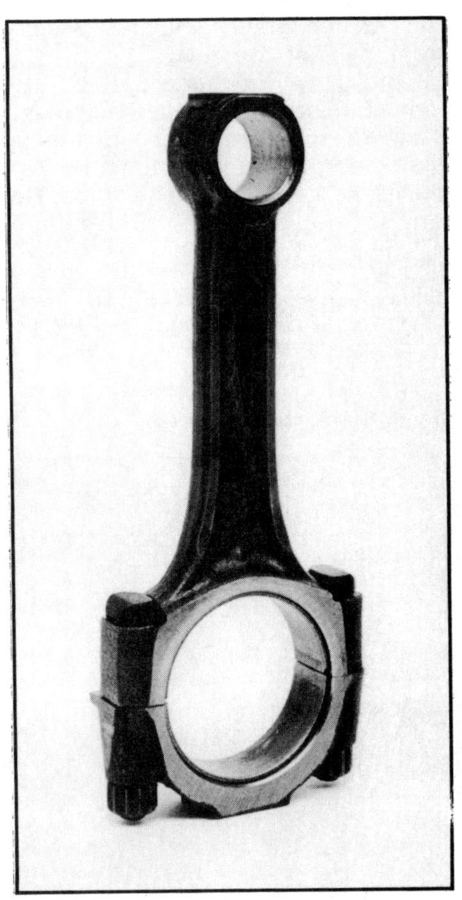

Production 60-degree rods closely resemble Chevy V-8 rods. They use 9mm metric through-bolts and nuts.

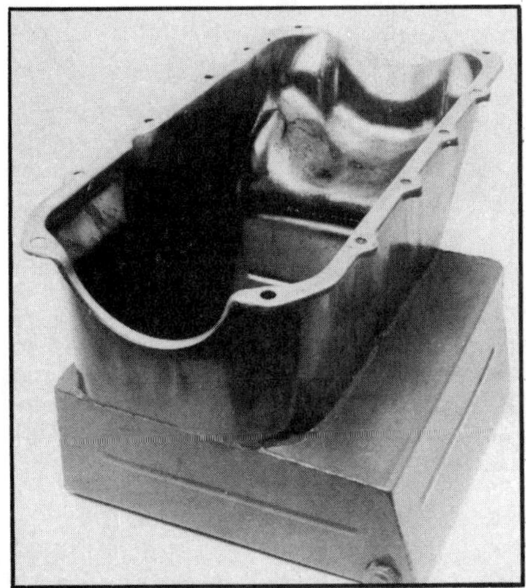

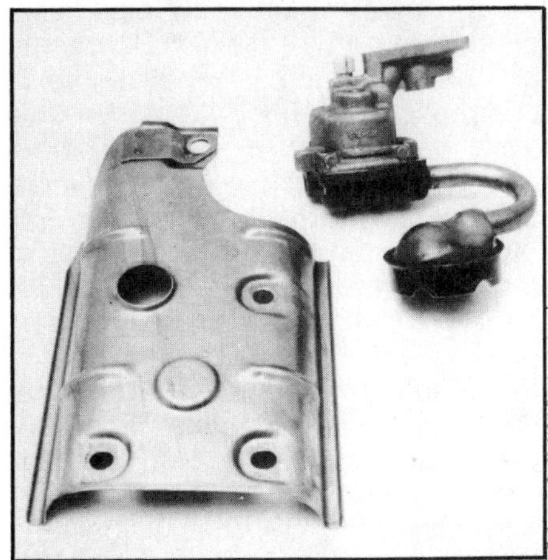

The Yenko Citation uses a stock pump with a cut down V-8 windage tray and an enlarged oil pan.

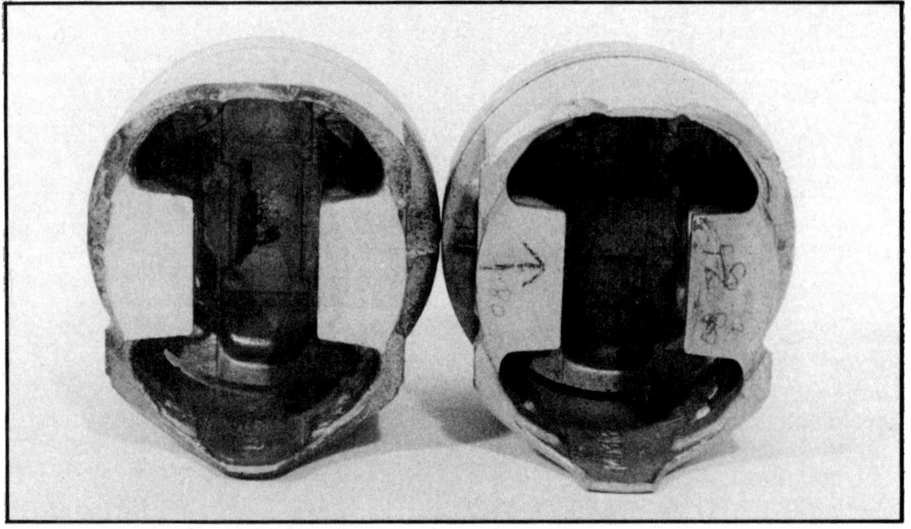

You really can't tell the difference between the standard piston (left) and the HO-660 version (right) just by looking. Both are aluminum flattops, but the high-output piston has approximately .020-inch taller deck.

each), and would seem to be the most accessible hop-up piston at this time for a typical street performance rebuild. However, the only significant difference between the H.O. and the production piston is an increase in compression height (the distance from the pin axis to the deck of the piston) of .020-inch. It might be more practical to mill the heads this amount rather than installing a new set of pistons, even though milling the 60-degree V-6 heads can cause problems (see cylinder head section). Other manufacturers, such as TRW, have not yet released performance pistons for this engine, but we expect they will appear in due time.

Considering the cost of the X-11 pistons, even street engine rebuilders should remember that any of the custom piston manufacturers (Arias, Venolia, Forgedtrue, etc.) can make

forged pistons for this engine. The advantage, besides the increased strength, is that the design can be custom tailored to your needs. The drawback, besides being even more expensive, is the waiting time and shipping delay necessary for custom orders.

For the original Yenko engine,

Katech used a set of Cosworth Vega pistons, which fit the V-6 block and rods. Since then, Chevy has developed a new Off-Road forged aluminum, dome-top piston, which appears to be an excellent piece for high-performance V-6 builders. Depending on the camshaft and valves used in the engine, it may have to be plunge-milled for valve reliefs, but there is enough dome volume to achieve as high as 11: or 12.5:1. For street applications, the dome could be machined down to lower the compression. This piston was still under development at press time, and hasn't yet been assigned a part number, but it should be available through Chevy dealers, and possibly TRW outlets, by late 1982.

CYLINDER HEADS

One big advantage of Chevrolet's 60-degree V-6 design over Ford's is that the cylinder heads have been designed to allow removal and installation of valve lifters without removing the heads. Consequently, a portion of the valve-cover sealing lip and pushrod clearance cavities are cast into the intake manifold, much like big-block Ford V-8's of the past. Since the stock intake manifold is aluminum, this design also eliminates weight from the cast-iron heads.

The stock V-6 heads, which are identical for right or left side, use rectangular intake ports, round exhaust ports, and an open-chamber design with a small squish area next to the valves. Stock valve sizes are 40.65mm (1.6 inches) intake and

Arias made a set of custom small-bore, 12:1, forged aluminum pistons for the MacPherson offroad S-10. The dome was cut by a tracer mill from a clay plug cast in the combustion chamber of the head. Stock rods are used.

33.2mm (1.3 inches) exhaust with 8.66mm (.341-inch) stem diameter. The exhaust valves are fitted with mechanical rotators, and the stock springs consist of an outer coil with a flat-wound inner damper.

As mentioned earlier, Chevy did considerable flow bench testing to maximize the breathing of these V-6 heads, increasing flow rates by 17% before they went into production. You will notice that each intake port has an odd-looking raised "bump" on the floor. This isn't some sort of smog control gimmick, so don't start grinding it away to hog out the ports. It was very carefully designed after extensive testing to effectively "raise" the port floor to improve intake mixture flow, especially at higher engine speeds. It's primary purpose is to help "bend" the incoming charge around the 90-degree corner between the port and the valve pocket.

Chevy's design plan for the 60-degree V6 was to work on the port design to maximize efficiency and power at higher rpm, then couple it with a relatively mild (short-duration, high-lift) camshaft to produce an engine that runs smoothly at lower speeds, has plenty of punch on the top end, and retains a strong torque curve throughout the operating range. Like typical Chevy engines, this one breathes well.

Whether cylinder head wizards will be able to find any more hidden horsepower in these ports remains to be seen. However, initial work by Diamond Racing, Katech Engines, and others seems to verify that Chevy did their job well. Extracting 250 to 300 horsepower from this V-6 requires virtually no port reworking other than the usual port matching and smoothing of casting irregularities. Using the stock intake manifold with a 2-barrel carb, the Katech-built Yenko motor produced 230 horsepower at 7200rpm on the dyno. With alcohol injectors, Shelley Arkow projects 300 horsepower from his circle-track V-6 with virtually stock H.O. heads.

Valves in the present head design are spaced far enough apart to allow room for insert seats should Chevy decide to offer aluminum head castings in the future. Consequently the iron heads can be opened up for bigger valves. However, the only larger valves now available are those used in the H.O. 660 option. The 1.72-inch intakes (part 14031328) and 1.425-inch exhausts (part 14031332) are 3mm (approximately .125-inch) larger in diameter in this engine. These valves can be installed in the regular heads by

Chevrolet will supposedly be releasing this dome-topped forged piston for the 60-degree V-6 sometime in the near future. Version at right has been flycut for additional valve-to-piston clearance with a high-lift cam.

The production 60-degree head comes with screw-in rocker studs and guide-plates. Note raised sections on intake port floors—described in text—and cutaway segment of valve cover rail.

Heads on the Yenko car were fitted with larger studs, Crane 1.5:1 Chevy roller rockers and Chevy dual valve springs (#330585) with performance valve seals.

Closeup of intake port in Yenko head gives a good view of bump on floor. As you can see, only very minor port grinding was done. Also note blockoff plate for manifold exhaust heat passage.

enlarging the pockets and regrinding the seats. You can also order the H.O. 660 headcastings (part 14031327), but they retail for over $160 each, bare, and the only difference between them and production heads are the larger valve openings. Combustion chamber volume is listed as 48cc in the production heads and 51cc in the H.O. heads—the difference being due to the shape of the valve heads, rather than the chambers.

Milling of the 60-degree V-6 heads to increase compression would not be practical for three reasons: (1) the thinwall castings won't allow much material removal, and warping could result, (2) the intake manifold sides

The MacPherson heads received considerably more port and polish work—by Ken Sperling of Air Flow Research (Sun Valley, California), including removal of much of the intake floor bump. These heads retain stock rockers, studs, guideplates and pushrods, which have shown no signs of failure when pushed beyond 7000rpm in competition.

and bottom would have to be machined to fit, and (3) a mismatch would still occur between the valve-cover rails on the intake manifold and cylinder head, as well as between the manifold and two vertical attaching studs on each head.

CAMSHAFT AND VALVETRAIN

Valvetrain design for the 60-degree engine is obviously patterned after Chevy's highly successful smallblock V-8 system. The six uses V-8 type lifters, tubular pushrods which feed oil to the top end, and stamped steel rockerarms mounted on individual studs with half-ball pivots and adjustable lock nuts. Stamped steel guide plates, held in place by the screw-in studs, align the pushrods and rockerarms. Stock rockerarm ratio is 1.5:1 and the 10mm screw-in studs are slightly larger than 3/8-inch in diameter.

Camshaft regrinds on the stock core are readily available from most aftermarket companies. The rest of the valvetrain pieces can be adapted from the huge variety of smallblock V-8 components available. V-8 solid or hydraulic lifters fit the V-6. However, since the main oil gallery intersects the left lifter bores, as on Buick and Chevy 90-degree V-6's, inertial-valve (often called "piddle valve") oil-control solid lifters should be used instead of the "edge-orifice" design. Since the lifter bosses are closer together on this engine than they are on V-8's (or 90-degree V-6's), no readily adaptable roller lifter now exists. Custom rollers for the 60-degree will have to incorporate both main gallery oil control and lifter aligning devices.

Chevy smallblock-type pushrods can be used in this engine if they are shortened to fit. Exact pushrod length will of course depend on valvetrain geometry with specific components used. Using a "cut to fit" V-8 pushrod, such as are available from Crane, would be the simplest to correct irregularities. The stock V-6 guideplates can be enlarged to accept large-diameter pushrods.

The V-6 heads can easily be drilled and retapped to accept Chevy V-8 screw-in rockerarm studs and your choice of aftermarket roller rockerarms. Standard 1.5:1 smallblock Chevy rockerarms do not fit the V-6 head perfectly, but come within .030-inch of lining up with the valvestem tips, which is very workable with roller rockerarms. By enlarging the valvespring pockets in the heads, most standard Chevy smallblock-type valve springs and retainers will fit this V-6 head.

Performance camshaft design for the 60-degree Chevy is still in the

Comparison of production head (top) with HO-660 head reveals that increase in valve head size is the only apparent difference.

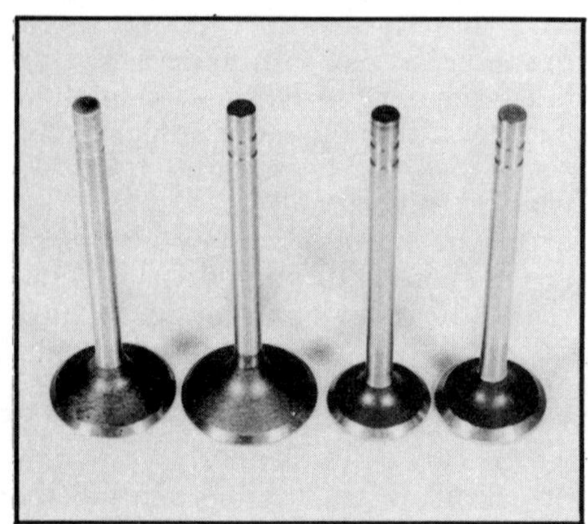

The HO-660 valves are 1/8-inch larger than production, and can be installed in the production heads if valve pockets and seats are correspondingly enlarged.

preliminary stages of development. As might be expected, most cam grinders are using Chevy smallblock V-8 profiles as a starting point. The Yenko Citation uses a General Kinetics cam patterned on the Chevy 3736097 off road V-8 cam, which features 220 degrees duration measured at .050-inch lifter rise, and .380-inch net valve lift. The MacPherson S-10 off-road pickup uses an Isky LL-505-T regrind with 242 degrees duration at .050-inch and .490-inch net lift. Shelley Arkow's midget engine is fitted with a Schneider flat-tappet "track grind" which has 258-degrees duration measured at .050-inch and .544-inch lift using 1.6:1 ratio Crane aluminum roller rockers (although Norris stainless steel roller rockers are shown in the photo of Arkow's valvetrain hardware, these would not fit during final as-

For his injected midget motor Shelley Arkow is using a Schneider #262F track grind cam, solid lifters and stainless steel roller rockers along with Chevy 7/16-inch rocker studs and high-performance V-8 pushrods, which will be cut down to fit the V-6.

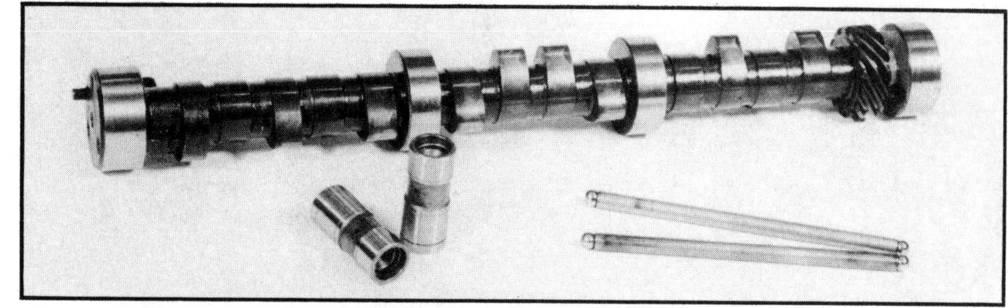

Nearly any solid or hydraulic performance profile can be cut on your stock cam by one of the aftermarket companies. Jerry McDonald is using an Isky 290° (242° at .050-inch), .490-inch lift cam with Chevy "piddle valve" solid lifters and the stock 5/16-inch pushrods.

sembly of the heads because of the close spacing between rocker studs).

For street modification, a good swap-in piece would be the hotter '81 X-11 hydraulic cam. Available from Chevy dealers under part 14031378 for about $80, the specs are 210 degrees (intake) and 217 degrees (exhaust) duration measured at .050-inch lifter rise with .393-inch and .410-inch lift, respectively. It could be installed without changing other stock components, though a set of aftermarket anti-pumpup lifters would be a wise addition.

To reduce both the size and the weight of the front cover assembly on the 60-degree V-6, Chevrolet lowered the position of the camshaft in the block and reduced both the diameter and the width of the timing gears and chain. Although the stock chain is only 3/8-inch wide, it is matched with hardened sintered-iron gears, and this assembly proved extremely reliable in prolonged durability tests during development. V-8-type timing gears will not fit on this engine, and no aftermarket roller-type chains have yet been introduced for it.

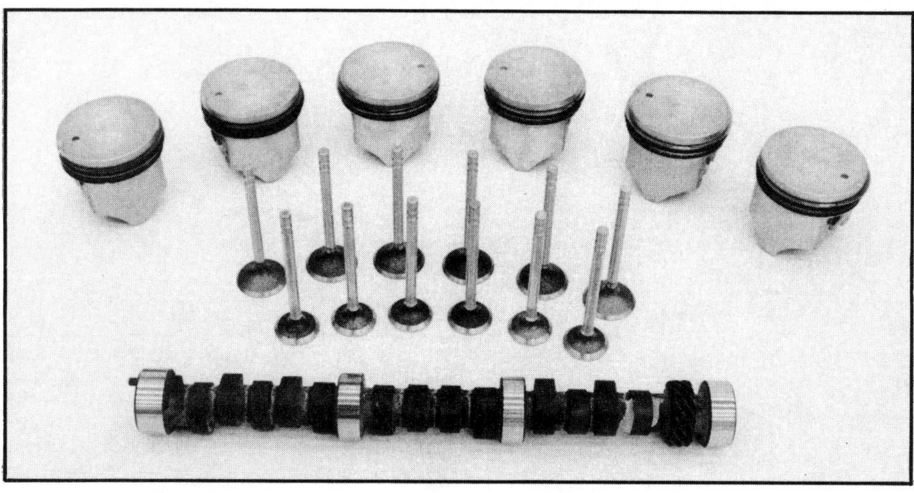

CHEVROLET 60° V-6 CAMS

Production camshaft 14024268

	INTAKE	EXHAUST
Opens	25° BTC	69° BBC
Closes	81° ABC	55° ATC
Overlap	80°	80°
Duration	286°	304°
Lift (net)	.347-inch	.394-inch
H.O.-660 camshaft 14031378		
Opens	31° BTC	83° BBC
Closes	93° ABC	61° ATC
Overlap	92°	92°
Duration	304°	324°
Lift (net)	.393-inch	.410-inch

(all figures measured at zero checking clearance)

Probably the most practical way to hop up a standard Chevy 60-degree V-6 for street performance is to upgrade it to HO-660 specs with the high-output pistons, valves, and cam, plus the X-11 ECM PROM for your computer box. If you don't want to go that far, the high-output cam by itself should be worth a few ponies.

A considerable amount of engineering went into the stock timing chain and hardened steel cam gear. It should perform very well in high-performance motors.

INTAKE, EXHAUST, IGNITION

Typically the first three targets for aftermarket performance products manufacturers, and the first three concerns of the "bolt-on" engine modifier—upgraded induction, exhaust, and ignition systems—currently do not exist for the 60-degree Chevy. And, unless the aftermarket devises methods for integrating, or otherwise coping with, the Computer Command Control system (or a large enough portion of the market is willing to discard the system) introduction of bolt-on manifolds, carburetors, or headers for this engine seems unlikely in the near future. This means the performance engine builder must modify stock parts, or else make his own components.

The stock intake manifold is a single-plane design, cast of aluminum, which mounts a 2-stage "Vara-jet" Rochester 2-barrel carb. With a small primary and a large secondary, this carb looks much like a Quadrajet sliced down the middle. All 1980 non-California versions use an air valve-operated secondary, while all others are completely controlled by the computer. Whether Chevrolet will switch to the Throttle-Body Injector (TBI) in the near future for this engine remains to be seen, but, because of distribution problems in the narrow manifold, there has also been talk at Chevrolet of computerized, direct-port injection for this V-6 in the future.

Of the stock intake manifold one Chevy engineer quipped, "That's not a manifold—that's a carb adapter!" One of the major drawbacks of any 60-degree V-6 design is the limited manifold space between cylinder banks. By extending the manifold into the normal cylinder head area, Chevy has improved the situation somewhat, but runner length and cylinder distribution remain less than optimum on this engine. The design of a really good

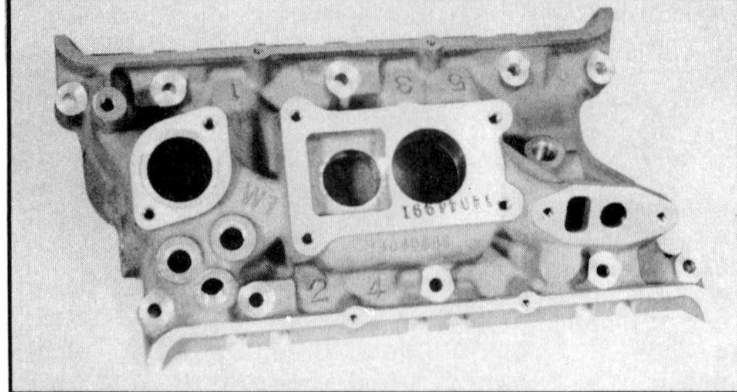

The stock, 2-stage, 2-barrel carburetor, shown in an S-10 pick-up installation (above), resembles a Quadrajet 4-barrel sectioned front to back. It has a large secondary. Also note the small body HEI distributor, with external coil, used on the S-10. Stock S-10/camaro manifold is shown at left.

The Yenko Citation manifold was modified to accept a bolt-on adapter plate for a 500cfm Holley performance 2-barrel. McDonald's intake was similarly modified, including extensive welding and grinding of ports and EGR passages, to accept the same type of carb. Note the difference between water necks on the Citation and S-10 manifolds.

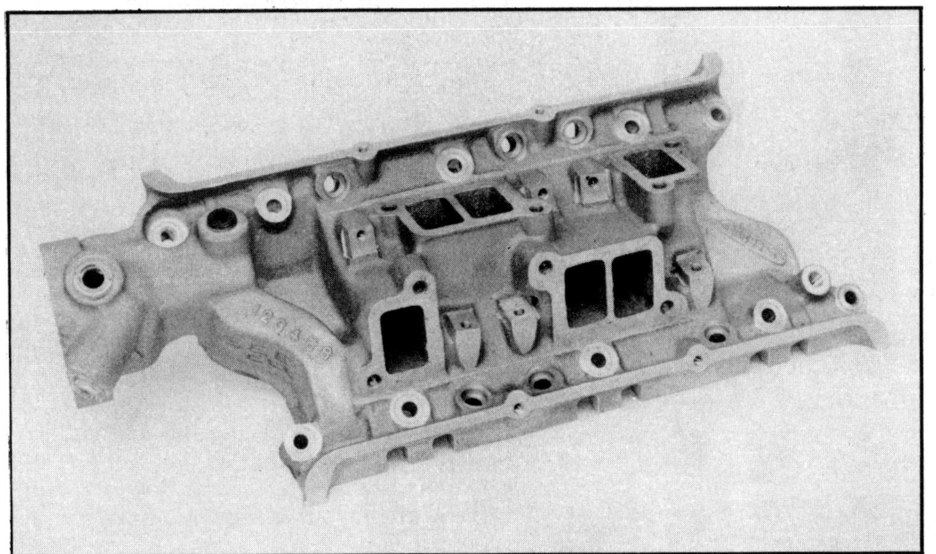

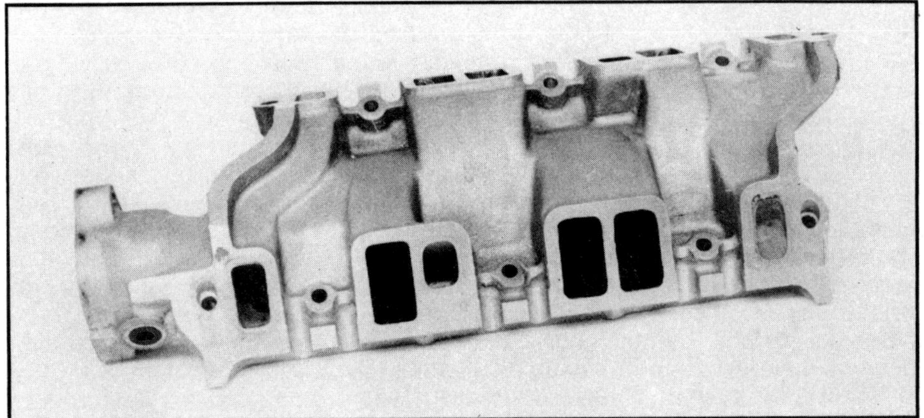

Shelley Arkow acquired this slick, one-of-a-kind intake casting, which was used during experimental testing of the V-6 with two 3-throat Weber carbs. This cast piece is basically an adapter base, to which various types of induction could be bolted (we may very well see this base used with Bosch fuel injection on production Chevy V-6s in the near future).

Arkow is using the manifold base shown above to adapt a set of Hilborn injectors originally made for a 215 aluminum Buick V-8, but cut down for the six.

intake manifold—possibly a hi-rise of some sort—will be a stout challenge for the aftermarket. Right now, it's the major missing link in the 60-degree performance package.

It appears, in the meantime, that modifying the stock intake manifold to accept some other type of 2-barrel carb will be the most practical route for modification. Both the Yenko Citation and the MacPherson S-10 use a fabricated aluminum plate adapted to the stock carburetor boss to accept a 500cfm Holley 2-barrel, as shown in accompanying photos. This setup has proven, while not perfect, at least very satisfactory.

More elaborate induction systems, such as a tunnel ram or direct-port fuel injection, pose a problem because the intake manifold incorporates a portion of the valve-cover rail and pushrod passages. Shelley Arkow was able to acquire a one-off Chevrolet aluminum casting used for a special testing session with a pair of three-throat Webers (from a 911 Porsche). To this base he is adapting a set of Hilborn injectors originally made for a Buick V-8. However, such a manifold is not available to the public. If you want to adapt injectors or other exotic induction to this engine, you will have to either cut up a stock manifold or machine some sort of adapter plates that will accommodate valve covers, pushrods, oil return, and water passages.

It should also be noted that the front-wheel-drive X-car V-6 manifold positions the water outlet and thermostat housing at the "rear" of the engine. If you intend to use this engine in a rear-wheel-drive application, you should be able to reverse the manifold, since the heads are identical, to place the water outlet at the front. However, the S-10 and Camaro use a different manifold with a front water outlet.

Stock exhaust manifolds are made of nodular iron to resist cracking and are very light in design. They could be easily hooked up to a free-flowing, dual exhaust system. Since individual tube-headers are not available, they would have to be custom made. Katech made their own to fit the Yenko X-body; Shelley Arkow had a set fabricated by Speed Fab of Canton, Ohio, using 1-5/8-inch primary tubing approximately 32 inches in length. Remember that this engine fires evenly on each bank, so all primary runners should be the same length, dumping into a common collector (about 3-inch o.d.). Katech also found that a crossover equalizer tube between the collectors dramatically increases low-end torque.

For now, you'll have to have headers custom built to fit the 60-degree V-6. Speed Fab of Canton, Ohio, made these typical sprint headers for Arkow's midget.

Non-computerized 60-degree engines come with a Delco HEI ignition system, and this is currently the only alternative for those who wish to dismantle the CCC system. Both the Yenko and Shelley Arkow cars use the stock distributor modified for full (locked) advance, though the Arkow car may be fitted with a custom-made, crank-trigger ignition in the future. The non-computer-controlled HEI may also be mated with aftermarket ignition boosters available from MSD, Accel, Stinger, and others. The S-10 pickup comes with a small-body HEI distributor which uses an external coil. However, it has no provision for mechanical advance since it is computer controlled. The MacPherson racing S-10 uses a custom distributor made by welding the V-6 base onto a standard Chevy six-cylinder distributor body, which is fitted with a Hays Stinger breakerless module. This modified distributor is firmly braced to the intake manifold to keep it from breaking due to engine vibration.

DRIVELINE

Front-wheel drive X-cars come with either a 3-speed Turbo Hydramatic (THM 125) or an overdrive, manual 4-speed (Muncie 125-4) matched to differential final drive ratios of 2.84:1 or 3.34:1, respectively. The front-drive powertrain seems to be relatively strong, and it is incredibly light and compact. The one weak link in the system so far, a byproduct of the severe downsizing of the assembly, has proven to be the manual clutch. Measuring only approximately 8.5 inches in diameter, it just doesn't have the pressure needed to hold a hopped-up V-6.

Unfortunately, the flywheel is of an unusual raised-surface design, making adaptation of any stronger clutch/pressure plate assembly difficult. It also demands that specific tooling be designed by aftermarket clutch manufacturers before they can offer something tougher for the X-car. One solution, being used successfully on the MacPherson S-10, is to add a ring of "centerforce" weights to the fingers of the stock diaphragm pressure plate. This unique modification, devised by Bill Hays of Midway Industries (Midway City, California—makers of Stinger Ignitions) increases the loading force of the pressure plate, and can be adapted to most diaphragm-type clutches.

Another drawback for performance enthusiasts is the extremely wide gear splits in the manual trans: 3.53, 1.95, 1.24, 0.81. That first to second gear

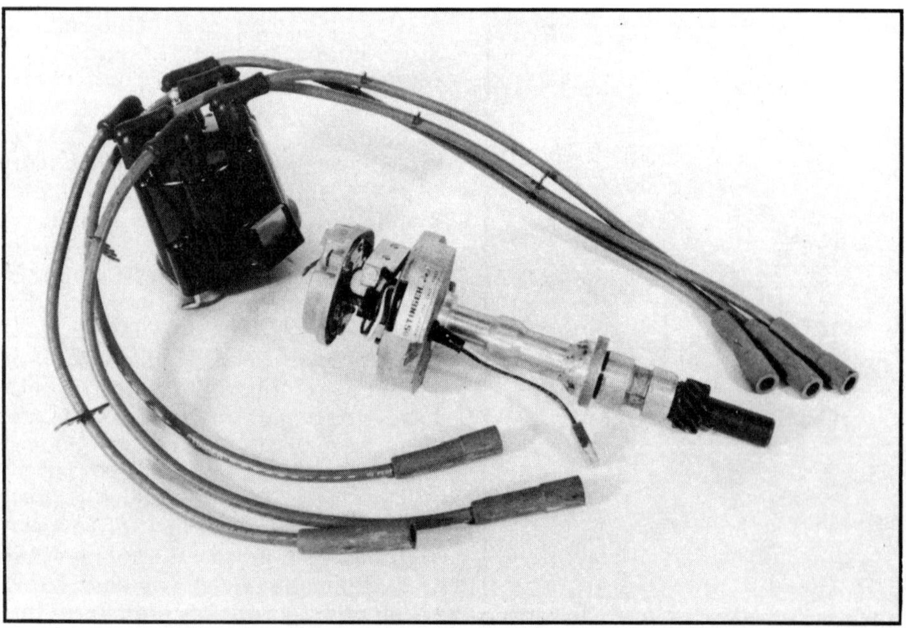

To make a simple, reliable ignition for the MacPherson S-10, Stinger welded the body of an early points-type GM six distributor onto the base of the S-10 HEI, then fitted it with their breakerless pickup and module. To keep the modified distributor from vibrating and breaking, especially on long offroad races, it is braced to the intake manifold with a strong bracket.

Although no competition flywheels or clutches are yet available for the 60-degree motor, Bill Hays of Midway Industries (same as Stinger Ignitions, Midway City, California) modified the stock diaphragm pressure plate in the MacPherson S-10 with his unique "Centerforce" weights on the clutch fingers.

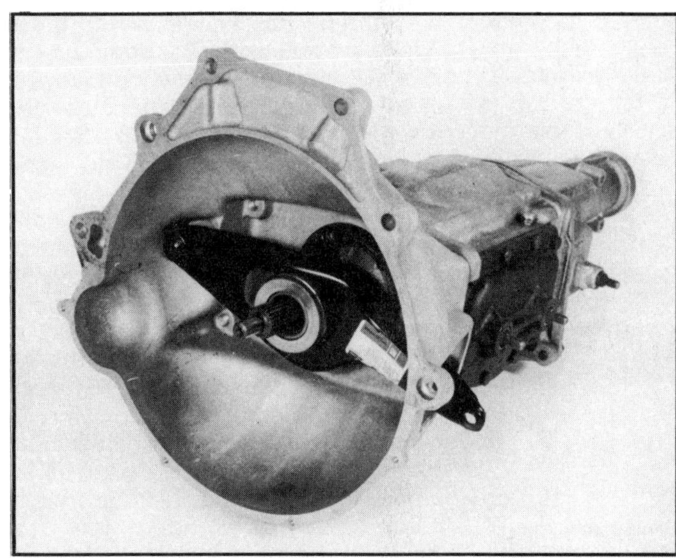

The S-10 bellhousing has a starter bulge on the right side and a long cable-actuated throw-out arm. It is shown here hooked up to a super T-10 4-speed.

The Camaro V-6 bellhousing features two starter bulges, for some reason, and uses a bellcrank throw-out arm common to Chevy V-8s.

jump is enormous. The Yenko Citation has been competing on the road course with a "secret" set of close-ratio gears specially cut for the X-car box—but such a modification is beyond the means of most enthusiasts. For straight line acceleration, the automatic 3-speed (with gear ratios of 2.84, 1.60, and 1.00) has shown slightly quicker times than the stick shift. Some models come with an overdrive automatic, however. If you're interested in drag racing an X-car, you could make a real tire-smoker by swapping the 3.34:1 "stick" differential gears in the automatic transaxle. Otherwise, parts to improve the X-car driveline just don't exist now.

In the rear-wheel-drive Camaro and S-10 options, a variety of transmissions have been used. In early 1982 the S-10 came with an Isuzu all-aluminum 4-speed with an integral bellhousing. In mid-'82 the S-10 was fitted with a Borg Warner 4-speed given the Chevy option code designation M-73. This aluminum transmission has a long tailshaft with the gear shift lever built in. It uses a separate aluminum bellhousing with a cable-actuated clutch fork. For late 1982 or 1983 a Borg Warner 5-speed overdrive transmission is also scheduled as an option, but hadn't been released by our press time.

The Camaro is available with a more traditional "side shift," cast iron Borg Warner or Saginaw 4-speed. It uses a separate aluminum bellhousing which has a rod-actuated clutch fork (similar to Chevy V-8's), and which also has starter "bulges" on both sides—which means this bellhousing could be used to convert an X-car V-6 to rear-wheel-

CHEVROLET 60° V-6 PRODUCTION SPECIFICATIONS

	Production	H.O.-660
Displacement	2.8L (173ci)	2.8L (173ci)
Bore	3.50-inches (89mm)	3.50-inches (89mm)
Stroke	3.00-inches (76mm)	3.00-inches (76mm)
Firing order	1-2-3-4-5-6	1-2-3-4-5-6
Cylinder numbering	1,3,5 right bank 2,4,6 left bank	1,3,5 right bank 2,4,6 left bank
Compression ratio	8.5:1	8.9:1
Main journal diameter	2.50-inches (63.35mm)	2.50-inches (63.35mm)
Rod journal diameter	2.00-inches (50.75mm)	2.00-inches (50.75mm)
Connecting rod length	5.70-inches (144.78mm)	5.70-inches (144.78mm)
Cam journal diameter	1.87-inch (47.46mm)	1.87-inch (47.46mm)
Valve stem diameter	0.340-inch (8.66mm)	0.340-inch (8.66mm)
Valve head diameter	Int: 1.600-inch (40.65mm) Exh: 1.30-inch (33.20mm)	Int: 1.725-inch (43.82mm) Exh: 1.425-inch (36.20mm)
Rockerarm ratio	1.5:1	1.5:1
Valve lift	Int: .347-inch Exh: .394-inch	Int: .393-inch Exh: .410-inch
Valve clearance	Zero	Zero
Main bearing clearance	.002-.003-inch	.002-.003-inch
Rod bearing clearance	.0015-.0025-inch	.0015-.0025-inch
Piston-to-bore clear.	.002-.003-inch	.002-.003-inch
Piston comp. height	1.60-inch (40.6mm)	1.80-inch (45.7mm)
Piston ring gap	.010-.020-inch	.010-.020-inch
	Bolt torque (pound feet)	Bolt torque (pound feet)
Cylinder head	70	70
Main caps	68	68
Rod caps	37	37
Intake manifold	22	22
Flywheel	50	50
Vibration damper	75	75

drive, since the X-car starter is located on the other side. Both the Camaro and the S-10 bellhousings use the standard Chevrolet transmission bolt pattern, and popular transmissions like the Muncie or Borg Warner T-10 4-speeds will fit them.

The only automatic transmission available so far in either the Camaro or the S-10 is the TurboHydro 200-C, which is a relatively weak three speed. The "C" designates that it uses a locking torque converter for improved mileage. Since the bellhousing flange is integral with the body of this transmission, and since the bellhousing bolt pattern of the 60-degree V-6 is unique, this is the only automatic which will bolt directly to this motor.

TURBOS, BLOWERS, NITROUS-OXIDE

Perhaps one practical way to modify a computerized 60-degree V-6 for significant power increases without dismantling the system would be to add a nitrous-oxide injection set-up. Kits haven't been developed specifically for this application, so you would have to have special parts, such as an under-carb adapter plate, made up by a nitrous equipment supplier, but most of the hardware is basically universal. Exactly what effect a shot of nitrous and extra gasoline will have on the exhaust system oxygen sensors, and how the computer will react, is difficult to predict. However, the nitrous system would operate independently of the computer, supplying its own gasoline and nitrous oxide when switched on, and since nitrous is an anti-detonant, it should keep the ignition timing from being retarded. However, we must emphasize that we are only speculating on the compatibility of nitrous injection and the CCC system. Before installing a system on a computer-controlled car, discuss the possibilities with a reputable nitrous oxide injection specialist. We would also suggest "testing" such a system cautiously, starting with a very mild dose of nitrous oxide and extra fuel, and working up to stronger jetting if everything appears to be working properly.

In lieu of other aftermarket speed products being introduced for this motor, turbocharging will likely be the most popular modification for non-computerized vehicles. Chevrolet designed a turbocharged 1980 X-11 Citation themselves as a one-off exhibit car. Using the standard RPO LE2 8.5:1 compression motor, they added a small Borg Warner/IHI RHB6 turbo with an integral wastegate set to five pounds boost. The blow-through setup pressurized a single throttle-body injector housed in a specially made aluminum box on the stock intake manifold. One of the most noteworthy aspects of this factory hot rod is that it *was* computer controlled. Delco engineers reprogrammed the ECM brainbox to operate the TBI and ignition, as well as a water/alcohol injector, in a mode compatible with the turbo.

Other turbocharged versions of the 2.8-liter have been scarce, and no kits have yet been introduced. John Anderson (AmTest Turbocharging Systems, Napa, California), one of several turbo experts who specialize in installing one-off applications, added a Rayjay EF-4 compressor to a Pontiac Phoenix for a Hot Rod Magazine test project (see the February '81 issue) with impressive results. Designing a draw-through system with a Carter 9500 series (500cfm) AFB 4-barrel, he

Chevy built this bright yellow turbocharged version of a 660 Citation as a design and engineering exercise. It actually has a production (non-high-output) version of the V-6, fitted with a small Japanese IHI RHB6 turbo set at a meager 5psi and blowing into an air box with a single GM throttle body injector. The package was dialed in by GM to work by computer, and it put out 170 horsepower at 4800rpm.

As proof positive that hot rodding is alive and well in the X-body age, check out Richard Meyers' 1980 Pontiac Phoenix sporting a 4-71 Jimmy and a 390cfm Holley on the 2.8-liter V-6. Richard adapted the blower and drive himself to the otherwise stock engine, which pulls 13-second quarter-mile times on 14 pounds of boost.

used 9.5 pounds of boost to increase the X-car's performance from 17.28/78.86 quarter mile times to 15.12 seconds at 93.84mph (with a slipping clutch in a 4-speed car).

Besides removing the stock carburetor and emission-control devices on the noncomputerized engine, no other modifications were performed. To control detonation, mechanical advance in the stock HEI distributor was reduced from 13 to 8 degrees by adding a small stop on the advance mechanism. A slot was then cut in the vacuum advance linkage to increase timing a like amount (but only under non-boost conditions). Initial static advance was set at 17 degrees for a total of 25 degrees. Using unleaded premium, the car reportedly had no detonation problems and was not fitted with a water injector.

So far we have seen only one example of a Roots-blown 2.8-liter V-6 Chevy. This striking 1980 Pontiac Phoenix, with a 4-71 GMC sticking sideways through the hood, was created by Richard Meyers of Minneapolis, Minnesota. Since he had access to a machine shop, the custom blower installation was relatively easy on the X-car. A flat aluminum plate was attached to the stock intake manifold to accept the supercharger, then a gilmer belt pulley was attached to the engine's crank pulley to drive it. The fact that the water neck on the X-car is at the back of the engine makes the installation of a blower much easier. Richard first used a set of drive pulleys to underdrive the blower 20%, but it still produced 14 pounds of boost in the small displacement motor. Since this was considerably more pressure than the stock pistons should be expected to bear, he subsequently installed a new gear set to underdrive the blower 40%. Other than the supercharger, the rest of the engine is completely stock. On 14 pounds of boost, it ran the quarter mile in 13 seconds.

The best news for the 60-degree Chevy V-6, however, is the new home it has found in S-10 pickups and new Camaros, providing traditional rear-drive mounting. This 220-hp example came out of Jerry McDonald's off road S-10. It is fitted with a Super T-10 four speed.

At the 1982 NHRA Winternationals Bill Maropulos' Chevy V-6 D/Econo Dragster walked away with the Competition Eliminator title, running so far under the index that the motor will probably be reclassified. He uses a sectioned V-8 tunnel ram with a single 4-barrel on the heavy-duty block and heads. (Photo by Les Lovett)

THE 90-DEGREE CHEVROLET

In an interview conducted for this book shortly after he had completed a testing session with the 90-degree Chevy V-6, Bill "Grumpy" Jenkins said this about the new Chevy engine: "The Chevy V-6 came along later than the Buick, and everybody thought the Buick was going to be the big hot dog. But as soon as Chevrolet decided they were going to do something, why, there's no way the Buick even has a chance." Jenkins may be somewhat opinionated in favor of Chevrolet, but he obviously has a good opinion of the new "smallblock" V-6. His test program, using the 4-inch bore, heavy-duty block, aluminum heads, and Chevy "cross-ram" manifold, netted 437 horsepower on the dyno. We will relay several of his interesting discoveries and suggestions as this chapter progresses.

Had both the Chevy and the Buick 90-degree V-6's been introduced at the same time, there's little question that performance enthusiasts would have turned to the Chevy first. The proven racing potential and the existing pool of parts and technology for the Chevy smallblock V-8, much of which applies directly to the cloned V-6, are impossible to ignore. But the Buick V-6, patterned after a far less

Soon after the release of heavy-duty components for the Chevy 90-degree V-6, Bill Jenkins screwed together a 265-inch motor and checked it out on his dyno. The master of the Mouse Motor found several expected similarities with the Chevy V-8 smallblock—and several unexpected differences.

famous V-8, preceded the Chevy by almost twenty years. Consequently, when the V-6 was "reborn" in the mid-'70's the aftermarket parts suppliers, the racers, and the new GM corporate mentality focused on the Buick.

Meanwhile Chevrolet, as usual, played the corporate maverick, rushing to get their own V-6 into production by 1978, possibly hoping to blow Buick in the weeds. But by then the question of technical superiority was a moot point. Hundreds of thousands of Buicks had been built, the aftermarket had tooled up, performance programs were being run, and the corporation was smiling. The Chevy 90-degree V-6 found its way into some Camaros and Chevy passenger cars for three years, but it was never offered in California. Meanwhile, Chevy engineering was busy designing the new 60-degree motor as their bid for a "corporate" V-6, apparently winning approval by 1982 (because it does a better job of meeting government standards). The 90-degree Chevy V-6 will be slated as the base engine in "G cars" (Malibu and El Camino), "B cars" (Caprice), and "G trucks" (vans) for the near future, but only the vagaries of federal and state regulations can determine for sure the extent to which this engine will be used in production vehicles.

What does all this mean for you, the performance enthusiasts? First of all, it means that not too many of you now own cars with these engines, and there won't be an overabundance of them in junkyards for others to buy. Consequently, the aftermarket will be somewhat hesitant to produce much in the way of typical bolt-on goodies specifically for this engine: manifolds, headers, ignitions, valve covers. It's not that such parts don't exist, but you won't see the variety of performance and dress-up components offered for the 90-degree V-6 Chevy that you do for more popular engines.

If Chevy isn't going to (or can't) drop the mini-mouse into new Camaros or Corvettes, what are they going to do with it? They're going to hot rod it, that's what. While certain sanctioning bodies such as NASCAR, USAC, and NHRA seem to be as vague about future rulings as the government is, Chevrolet has been building parts for this engine as if they intended for it to win the Indy 500. So if you're a serious performance enthusiast, there's plenty of hope for the 90-degree Chevy. In fact, of the four engines covered in this book, it appears to have the most brute performance potential of all.

We're talking about serious racing or off highway use. Though you may not find much in the junkyards or parts stores for mild street performance, the basic components for a heavy-hitter V-6 are available from Chevrolet: a 4-bolt main, 4-inch bore, heavy-duty block; aluminum, big-valve heads; performance cams; and even a radical "rat-roaster-type" intake manifold.

Given these necessary ingredients, almost any other high-performance parts you might need—pistons, rods, valves, lifters, oil pump, water pump, you name it—can be supplied either by Chevrolet or the aftermarket from the vast supply of interchangeable smallblock Chevy V-8 compo-

nents. The initial 200 cubic inch (3.3-liter) 90-degree V-6 ('78-79) was essentially a 267-inch V-8 with cylinders 3 and 6 sliced out of the middle; they share the same bore and stroke of 3.50 inches by 3.48 inches. The '80 and later 229-inch version has a bore increase to 3.74 inches—the same as the 305 V-8. All of these engines share with the 350-inch smallblock V-8 the same stroke, rod length (5.70 inches), rod journal diameter (2.10 inches), and main-bearing journal diameter (2.45 inches). The 4-inch bore heavy-duty V-6 block of course shares the same cylinder bore size as the 350, 327, and early Z-28 302 smallblock V-8's. The list of V-8 components that will swap directly into the V-6 is amazing: valves, springs, rockerarms, pushrods, timing chain and sprockets, lifters, camshaft bearings, main bearings, pistons, oil pump, front cover, water pump, harmonic balancer, flywheel, bellhousing, starter...and more.

Obviously much of the approach and specific procedures for modifying this V-6 translate directly from well-known V-8 technology. Numerous books and articles have been devoted to subjects like smallblock valve train design, port configuration, piston and combustion chamber shaping, and so on. The scope of this book doesn't allow a thorough investigation of all the latest smallblock Chevy performance tricks that might apply to the V-6. Such information is readily available elsewhere, such as *The Chevrolet Racing Engine* by Bill Jenkins and *Chevy Performance* both available from S-A Design Books (see page 128).

Instead, we will address ourselves more to the *differences* in the mini-mouse. One of the pitfalls of working on this engine is relying too heavily on

past V-8 experience. As Grumpy Jenkins put it: "There are a lot of details that don't come out right—the pushrod locations, the lifter locations, and things like that. ...There are many little physical differences that you're not aware of, but you walk up to it and all of a sudden it gets you. You're ready to bolt something together and it turns out not to be the same."

Such dimensional inconsistencies are minor annoyances for builders

The 90-degree Chevy can be built as radical as any smallblock—even for street. This BDS-blown and injected bent-six amply powers Car Craft magazine's Pro Street project Malibu.

used to working on smallblock V-8's. Differences of far greater concern to the performance engine builder are the V-6's "split," narrowed, and offset connecting-rod journals, the uneven overall firing pattern (108°, 132°, 108°, etc., crankshaft rotation between ignition pulses), and the 1, 6, 5, 4, 3, 2 firing order which produces an even firing sequence in each separate cylinder bank. We will explore these peculiarities of the Chevy 90-degree V-6, as well as V-8 performance part interchangeability and aftermarket part availability, in the following discussion.

BLOCK AND OIL SYSTEM

So far three different cylinder blocks have been offered for the Chevy 90-degree V-6. The '78-79 200-inch production version has a 3.5-inch bore, while the '80 and later 229-inch production block has a 3.74-inch bore. Otherwise these two blocks are identical, both being lightweight, thinwall castings with 2-bolt mains. The earlier model cannot be bored out to 3.74 inches (it is cast with different cores for the smaller bore). In fact, neither of these production blocks should be bored much larger than .040-inch.

The production blocks are very similar to the late-model Chevy V-8 from which they were derived. Most basic dimensions, machining angles, accessory mounting bosses (oil pump, distributor, front cover, water pump, bellhousing), and some components (e.g., main-bearing caps) are identical with the 267-305-350 V-8 blocks. The main differences, besides the removal of two middle cylinders, are "extensive internal revisions to obtain the maximum weight reduction possible"

(quoting the Chevy V-6 SAE paper), and a basic change in the oiling system, primarily for the same reason. The production V-6 design eliminates the central main oil gallery, cast in the middle of the lifter valley on V-8's. In its place are lightweight cast-in "tie bars" between the lifter bosses. The right-hand lifter oil gallery remains the same as the V-8's, but the left-hand gallery is offset and is increased from 7/16-inch to 9/16-inch. It serves as the main oil gallery feeding the camshaft, crankshaft, and rod bearings, in addition to feeding the left-hand lifters and valvetrain. Although the oil pump remains in the V-8 location, oil passages in the block were rerouted (as shown in the accompanying diagram) to affect this change. Further alterations to the

production V-6 oil system include the substitution of a smaller 70mm oil filter with a cast-in screw-on mount and the oil bypass valve is relocated to a passage in the block.

Another minor difference between the V-6 and the V-8 is the spacing between lifter bores. Although both engines use the same size and design of lifters, the distance between lifter bore centerlines on the V-6 is 1.56 inches for the front and back cylinders, and 1.67 inches for the middle two cylinders.

As we mentioned, you're not going

to find an abundance of these production V-6's available (compared to smallblock V-8's, for instance). But they're not exactly scarce, either. If you can find one, or if you have one in your car, it's excellent material for a mini-mouse street motor which will respond to typical street-type modifications.

For more serious engine building, however, we would strongly recommend the heavy-duty V-6 block. This version is cast like the 400-inch V-8 smallblock, with thicker coring all around, beefier main bearing webs, and a lifter valley cast (though not drilled) like the V-8. This block has never been offered in a production vehicle—it's a special-order item from Chevrolet, currently listed in two bore sizes: 3.75-inch (part 14014441) or 4-

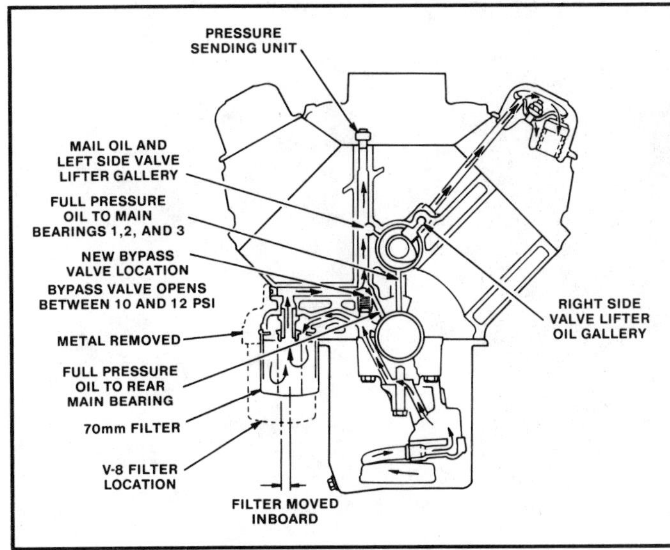

The oiling system in the 90-degree V-6 differs significantly from that in the V-8. Like the Buick V-6, it uses a lifter gallery as the main gallery. Oiling is the same in production and heavy-duty blocks.

The heavy-duty (special-order) block is cast just like a 400-inch V-8, with siamesed cylinders and a closed lifter galley (not drilled for V-8 type oiling). It is available in 3-3/4- or 4-inch bore sizes.

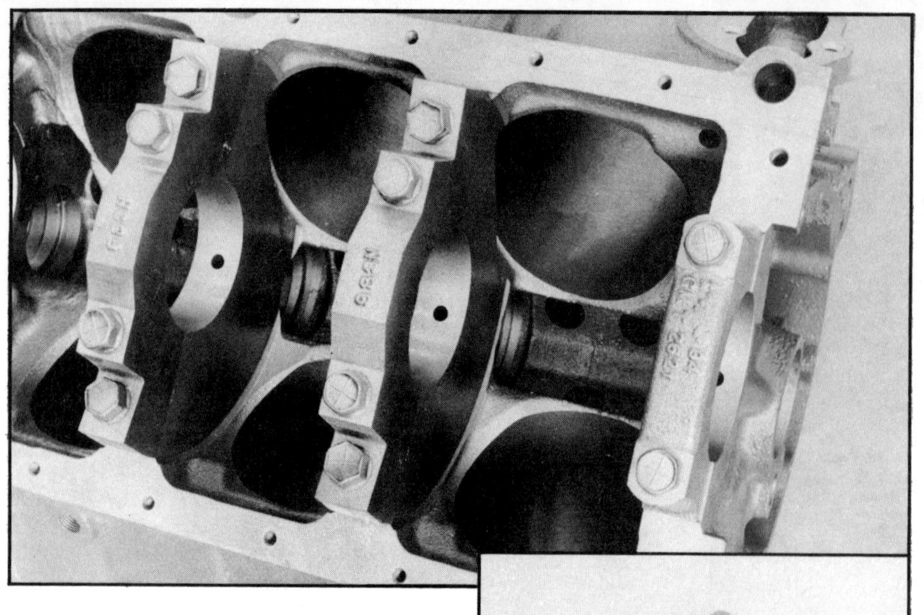

Substantially thicker main bearing webs are evident on the heavy-duty block. Although it comes with standard main caps, it can be easily fitted with 4-bolt mains, available from Chevy (inset) or aftermarket suppliers.

inch (part 14011066). These blocks come with V-8-type, 2-bolt main caps, but will accept Chevy (part 14011039) or aftermarket 4-bolt main caps simply by drilling and tapping holes for the outer bolts and align boring the assembly. Like the 400 V-8, this block has siamesed cylinders, and they have enough cylinder wall thickness to allow overbores up to .125-inch.

Although the heavy-duty V-6 block is cast with a central valley chamber runner as used in the V-8's for a separate main oil gallery, this block comes with oil passages drilled like those in the production V-6. That is, the central runner is blank and the left lifter gallery is drilled oversize and acts as the main oil feed. As noted in the 4th edition of the *Chevrolet Power* manual (which devotes a section to the new V-6): "At this time only the stock V-6 oiling system is provided, and deemed to be adequate. However, the block can be drilled and converted to the V-8 oiling if desired by the customer." Builders who have explored this engine for all-out racing, such as Bill Jenkins and Ryan Falconer (builder of the turbo-charged Chevy V-6 for Indianapolis in 1980), have stuck with the V-6 oiling pretty much as is. Jenkins, for one, felt that redrilling the block for V-8 type oiling would not be worth the effort. He put it this way: "As long as you have the right lifters and you run enough oil pressure, I just don't see the value in doing that. It's a lot of work to convert over to the V-8 system. You can do it, but you have your main feed from down

at the filter that you have to block off and reroute and so forth to get it hooked together. And if you're going to convert an already machined block, you'll still have one gallery that will take the funny lifters (or you would have to sleeve each left-hand lifter bore), and then you have to get the feed into it going in the opposite direction from the back. It can be done, but it's not easy."

For his testing program Jenkins fitted the engine with a dry-sump system (special Moroso pan, two scavenge pumps, one delivery pump running 80 pounds pressure, remote filter and cooler), but did very little to the passages in the block. "There's a bad entry location out of the filter that needed to get opened up and radiused...where the oil under pressure re-enters the block and filter boss there's a bad restriction at that point. That's about the only place we did anything to the oil system." Bill used the Crane oil-control V-6 roller lifters, which we will discuss in the valvetrain section.

Ryan Falconer redrilled a heavy-duty block for V-8 type oiling on his initial Indy test engine, but went back to the stock V-6 setup on the engine that actually ran the race. He also used a dry-sump system, but did "virtually

Jenkins used Chevy 4-bolt mains on his test motors, and dry-sump oiling. The only other change to the block was the addition of a small machined tab (inset) to adapt a big block starter, allowing the use of a larger flywheel. Note bowtie and "HD" emblem on side of block.

nothing, other than normal clean up of the passages" for the oil system in this motor. He also used the Crane roller lifters.

We talked with Falconer just before this book went to press and he was at that time working on two *all-aluminum* Chevy V-6's for the '82 season. The blocks were supplied by Chevrolet, cast in the same configuration as the heavy-duty iron block. Aluminum V-6 blocks have not been made available to the public yet, but Chevrolet has certainly demonstrated a good possibility that they might in the future.

For 1983 Chevrolet is planning to introduce a 262-inch version of the 90-degree V-6 as the base engine for Chevy vans. This motor would come with a 4-inch bore, but it would use a new casting of the current standard-duty block, not the heavy-duty, siamese-cylinder V-6 block. If this engine does go into production, it will supposedly come equipped with an aluminum, hi-rise 4-barrel intake manifold and Quadrajet carb.

Finally, as a point of interest, we should mention the "killer" V-6 lurking in the wings. Currently undergoing a long-term development process at Yenko Chevrolet (McMurray, Pennsylvania), it will be based on an adaptation of their existing all-aluminum rat motor block. At this point the rat V-6 exists only as a prototype block made by *literally* cutting two cylinders out of the aluminum big block and welding it back together. However, casting tooling is supposedly now in progress, with development of compatible heads and crankshaft to follow.

CRANKSHAFT AND CONNECTING RODS

The most significant difference between the 90-degree Chevy V-6 and its parent V-8 is the V-6's staggered-throw crankshaft. However, unlike the even-fire Buick which has 30-degree offset pins, the Chevy is a compromise between the even-fire and full odd-fire (paired pins, as in the V-8) configurations. It uses an 18-degree pin offset, and a consequent 108°-132° odd-firing sequence, arrived at through a random testing process described in the introduction to this book. Further, the reciprocating assembly is slightly underbalanced (at about 46%) to reduce the vertical primary unbalance vector. By reducing the offset, and thus increasing cross-sectional area between adjacent pins, Chevy could use standard V-8 pin diameter (2.10 inches) without substantially weakening the crank. With a

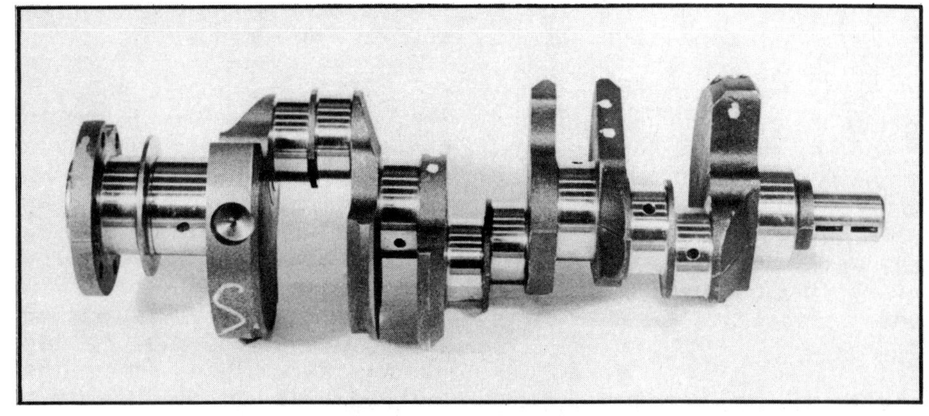

The Chevy V-6 crank resembles the Buick even-fire in its split throw design. However, the Chevy is still an odd-fire motor because the rod journals are split only 18 degrees. Both the 200 and the 229 use the same 3.48-inch stroke crank.

.180-inch flange between pins, journal width was reduced to .900-inch (.130-inch narrower than the V-8), resulting in rod journal centerlines that are offset .100-inch from the bore centerline.

There seems to be little problem with the integrity of the stock cast nodular-iron crankshaft for street or mild performance use. The same piece is used in both the 200- and 229-inch engines, and it's the only crank made by Chevy for this V-6. Naturally, this crank also fits the heavy-duty blocks.

For moderate performance applications, the stock crankshaft should receive the usual prep of Magnafluxing, oil hole chamfering, journal polishing, and Tufftriding (or similar processes). If the engine will be used for competition, or if you don't mind a bit of vibration in your street machine, have the assembly balanced at 50% of reciprocating weight to reduce overall unbalance to a minimum. The Chevy Power manual recommends cross drilling the main journals if the stock crank is to be used in competition, and shows diagrams for this operation. However, for most types of competition where the stock crank is applicable (or adequate), such a step is probably unnecessary. When installing the crank, use top quality V-8-type main bearings, such as TRW CL-77 (part MS 3454P). TRW has also recently released Clevite 77 rod bearings to fit the stock V-6 rods (part CB 1227P).

For any serious racing with the 90-degree Chevy V-6, such as long-distance, high-rpm oval or road racing, the builder will want to use a steel billet crank for two reasons. First, of course, it offers far greater strength and reliability. Secondly, the billet manufacturing process allows considerable freedom to redesign or relocate the crank throws. Racing engine builders are not so much concerned with the degree of crankpin split, but rather with the side effects of split-journal design—the narrowing and the offsetting (from cylinder bore centerline) of the rod journals. The stock design, with most of the offset at the connecting rod big end, obviously places greater loads on one side of the bearing than the other. The fact that the bearing width has been reduced only com-

Cutting a steel billet crankshaft for the Chevy V-6 not only gives a stronger part, but also allows the engine builder to change stroke length, piston timing (degree of rod journal split) and rod journal width and offset.

pounds the problem. The V-6 crank creates a classic dilemma for the engine builder: if he centers the rod in the piston, it offsets the big end, side-loading the bearing; if he centers the rod on the crank, it offsets the little end in the piston, creating hazardous side loads on the piston skirt and cylinder wall; if he goes to a shared crank pin (as in the V-8), he eliminates the rod offset problem but introduces the vibration and ignition problems of the 90-150° odd-fire configuration.

Bill Jenkins, in our interview, was one of the first to discuss these problems and possible solutions. Here are some excerpts from that discussion:

Jenkins: It seems to be mandatory to eliminate the flange (between rod journals) as much as possible.

Question: You used one billet crank with V-8-width bearing journals, but these journals are still offset from the bore centerline somewhat?

Jenkins: Oh yes, it's offset quite a bit still. But it uses the V-8 width, and you can use the V-8 bearing. All you have to do is chamfer the edges on both sides, but you still maintain the width of the bearing in the steel part, where the strength is. The V-6 bearings won't stay straight on rods—too skinny. They'll start to shift in the rods and edge load on you. You have to get rid of the flange as much as possible and get the rod over in the middle of the piston. And you have to stabilize the rod on the crankshaft—you can't run a lot of endplay. Possibly down to .007-inch or maybe even .005-inch, and grind some oil splash reliefs in the rods in a point where they're not loaded. Try to keep the rods perpendicular to the crankshaft. The load isn't straightened out through the rod and it tries to whip over to the side at the top and it will slap the

piston piers.

Question: What have you considered for solutions?

Jenkins: We had some ideas about putting rings around the dog pin like Kawasakis do or pressing the pins with buttons in them and getting the pin to hold the rod straight up and down. However, using pressed pins would be a very tedious operation. You'd have to press the pin in and move it in very small increments until you got it close enough to the middle of the cylinder and then finish up your centering of the pin in the cylinder with washers behind the buttons. It would end up being a button pin to maintain the rod position in the piston, instead of side loading the piston with spacers on the pin like the Kawasakis do. Whether or not offsetting the big end would be better than the pressed pin thing to stabilize the top of the rod, or to angle the beam, or something, so that it went basically from the very middle of the piston to the very middle of the crank pin, I don't know. The cylinder walls don't like to have a lot of end force on the piston, against the walls, they just don't like it. They'll split the walls. There's an awful lot of side force in the connecting rod because of the offset. You can get the beams in the middle of the pistons, and offset the rods at the bottom, but you're still going to get a certain amount of vector force away from each other. You'll take all the moment out of the top of the rod if you put the beam in the middle of the piston, but you will create a little bit of moment in the bottom of the rod, from the bearing being so far offset from the center of the beam.

Question: Do you feel there is any problem of weakening the crank by narrowing the flange between crank pins?

Jenkins: No.

Question: Even if it's not a billet?

Jenkins: Not really. The Buicks run down to .060- .080-inch (flange width), or something like that, and they don't have any problem on an iron crank, and they have more pin offset (splay angle) than the Chevy.

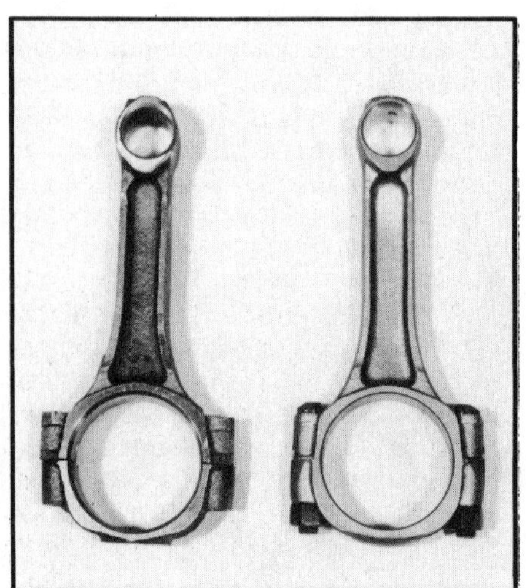

Close-up of a stock and a billet crank shows narrow flange on the latter, allowing wider rod bearing surface and less rod offset in the bore.

The stock V-6 connecting rod, shown in production form (left) and after polishing and shot-peening (right) for greater strength, should work fine in most moderate performance build-ups.

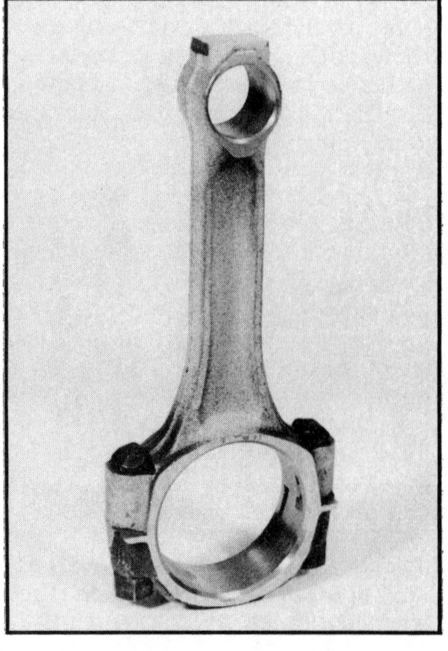

Chevy factory high-performance rods, such as those used in Z-28 and LT-1 smallblocks, can be swapped into the V-6 after the big end has been narrowed and chamfered on both sides. Using a V-8 rod will also allow narrowing of the rod journal flange (widening of the bearing surface) on the stock V-6 crank.

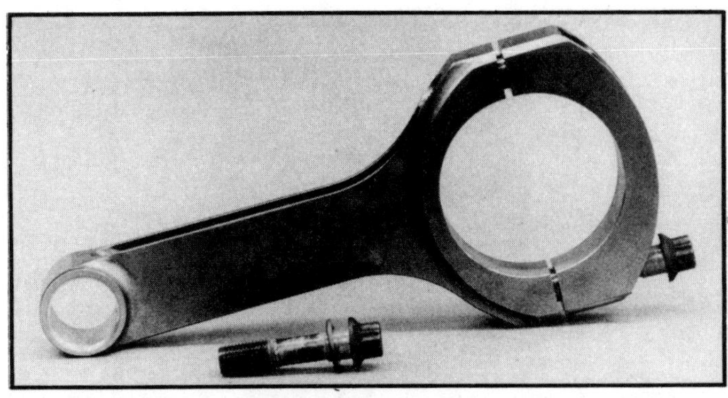

Having custom rods made for the V-6 allows plenty of freedom to play with bearing width and offset. Note how this Carrillo steel rod has been offset on the big end (right photo).

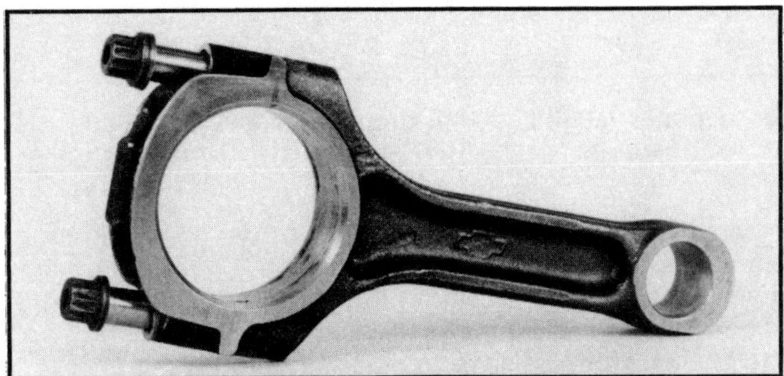

Chevy also now offers their own forged steel "bowtie" connecting rod, which can be used in the V-6—if altered in similar fashion to any other V-8 rod fitted to a V-6.

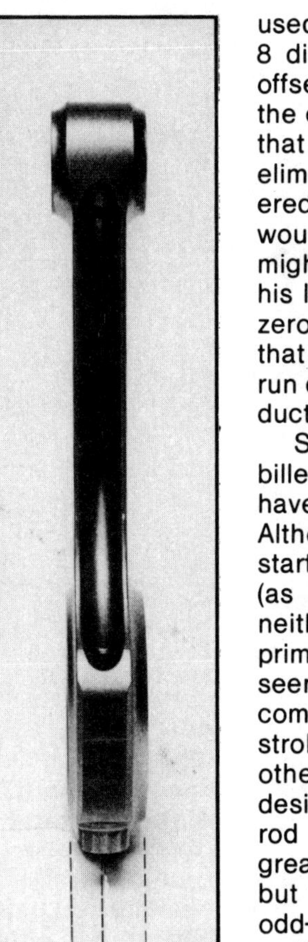

Question: If you were going to build a billet crank for this engine, would you use the stock offset, go to even fire (30°), or use the V-8-type pins?

Jenkins: If you're going to build a big motor, you can't do anything but go no offset (use V-8 type common pins). If you're going to go 3.75-inch stroke, there's no way this thing is going to stay together unless you use a common pin. Then you're down to 90-150° firing order, and it doesn't do anything for the ignition. The only way you're going to get out of the ignition scam is to go 3.25-inch stroke or less and go with a 30-degree offset (even-fire). Then you have no problem with the ignition and less vibration. I don't have any experience driving the darn thing so I don't know what the 90-150° common crank pin feels like, but I've had comments from two different drivers saying the vibration makes your arms feel numb in a long-distance race. This wouldn't be a consideration in a drag motor. If you wanted to build a big drag motor, you'd just build a common crank pin—but you're still in ignition trouble.

Obviously there aren't any easy answers to this complicated situation, but engine builders have just begun to work on it, and it will be very interesting to see how this aspect of V-6 tech-

nology develops. It should also be emphasized that Jenkins was working on a very specific program—relatively big-inch motors for sustained, high-rpm, road racing—and that he likes to test engines to their limits.

The rod bearing and rod stability problem seems to be much less severe in a short-stroke motor, such as the 209-inch turbocharged "stock-block" V-6 built by Ryan Falconer for Indy. Using a Moldex billet crank cut in the stock 18-degree offset configuration, he had the crank pin flanges narrowed to .125-inch and the journals widened to approximately .916-inch. To reduce piston weight (by increasing pin height) in the 2.75-inch stroke motor, he used longer (6.135-inch) Chevy L88 big-block connecting rods (part 3969804), narrowed slightly on the "inboard" side of the big end. Since the big-block rod has a slightly larger big end bearing diameter, the crank pins were enlarged to 2.200 inches on the billet, thereby further adding to the strength of the crank. Falconer pointed out that the same engine was used for practice, qualifying, and completing the race with no problems. However, at Michigan later in the season the car was sidelined with a rod failure.

For a 4-liter (244 cubic inch), normally-aspirated Indy V-6, Falconer

used Carrillo steel rods made to 350 V-8 dimensions, but with the big end offset. On the question of narrowing the crank pin flange, Falconer stated that he felt that it could possibly be eliminated entirely, even with staggered throws—although such a move would certainly weaken the crank and might lead to other complications. For his latest Indy V-6's, he plans to use zero-offset, V-8-type crank pins, noting that these full-odd-fire motors seem to run quite smoothly in dyno tests conducted so far.

So if you are going to have a steel billet crank made for a Chevy V-6, you have several design options available. Although most builders so far have started with the 18-degree pin offset (as in the stock engine), which is neither even firing nor balanced for primary reciprocating forces, it would seem most logical to go to the V-8-type common pin on a billet crank for a long stroke V-6, especially for drag racing or other short-duration contests. Such a design would eliminate the off-center rod problem, as well as afford the greatest cross-sectional pin strength; but you would have to contend with odd-fire vibration and ignition timing.

If you choose to retain the standard crank configuration, the hot tip is to narrow the flange considerably, and widen the rod bearing journal. As Jenkins pointed out, this procedure applies to the stock cast crank as well as to billets. Then switch to V-8 rods, narrowing one side of the big end to compensate for the specific flange width remaining.

Naturally the variety of aftermarket connecting rods adaptable to the 90-degree V-6 is nearly limitless, since just about anything is available for smallblock Chevy V-8's. If you want to experiment with big or little end offsets, most companies, such as Carrillo, can make up what you need from V-8 blanks. For less radical engines, you can choose from two excellent Chevy rods to adapt to the stock (or modified) V-6 crank. Both the Z-28 Camaro (part 3946841) and the LT-1 Corvette (part 3973386) 350 V-8 rods have the same 5.7-inch center-to-center length, 2.1-inch rod-journal diameter, and .927-inch wrist-pin diameter as the V-6. However, the stock V-6 rods are .050-inch narrower on the big end, and use a bearing that is .130-inch narrower than the V-8's. Stock V-6 rod bearing width is .708-inch, nominal. (Remember that V-8 rods, which are paired on the journal, ride on the crank pin fillet only on the "outboard" side; they ride against each other on the sides facing inboard. The V-6 rod must be cham-

The stock 229 piston is a slightly dished aluminum casting.

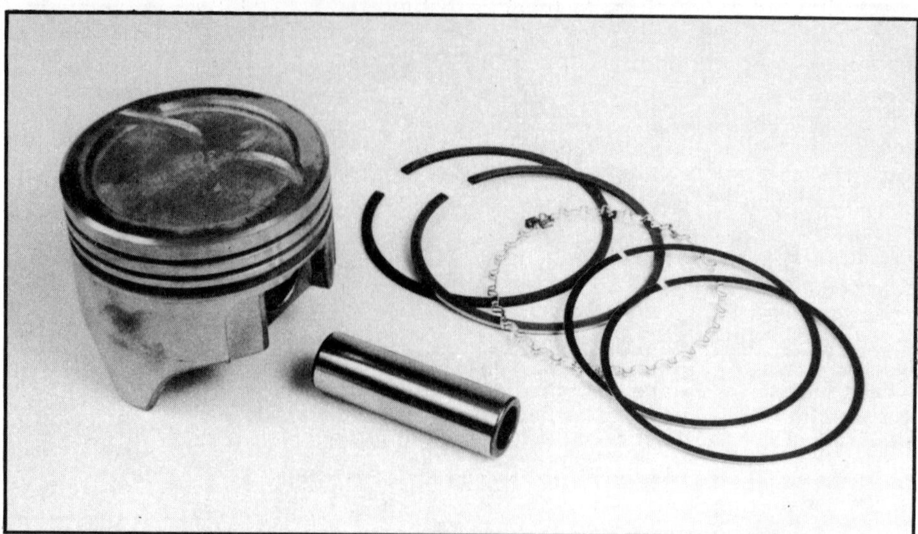

Since the 229 V-6 uses a piston similar to the 305 V-8, a couple of suitable aftermarket forged pistons (designed for the 305) are available from companies such as TRW.

fered for clearance on both sides, and the bearing must be narrower and chamfered on both sides.) To adapt a V-8 rod to the stock V-6 crank, narrow the big end approximately .050-inch on the "inboard" side (to approximately .890-inch), cut a .0625-inch chamfer on the same side, and install the V-6 bearings, being sure that they are centered in the rod so they won't ride up on the fillet radius. If you are going to widen the rod journals on the stock crank by narrowing the flanges, then you must adjust the figures given above accordingly, and you will have to narrow and chamfer V-8 rod bearings to fit. If you are adapting production V-8 rods to the V-6, remember that full-width V-8 bearings and journals can-

not be used, since this would require either cutting away the crank flange completely, or offsetting the top of the rod considerably in the piston, neither of which is recommended with a stock cast crank. To use the V-8 width bearings, you must use custom aftermarket rods with offset big ends.

Chevrolet has also recently introduced their super-trick, forged steel "bow tie" rod in both 5.7-inch (standard smallblock length) and 6.0-inch lengths. The shorter rod (part 14011082 for polished, shot-peened, and Magnafluxed version; 14011090 for "rough" version) can be adapted to the V-6 as outlined above.

The tolerances given here for modifying V-8 rods to fit the V-6 crank are

merely guidelines. As Jenkins noted, very tight rod side clearances might work the best. If you widen the crank pin, or move the rod beam, you will have to machine the rod big end, and chamfer or narrow V-8 rod bearings accordingly to fit properly in the specific engine you are building.

PISTONS

There is virtually nothing to report, at this time, in the way of over-the-counter performance pistons for the 200- or 229-inch V-6's. The production engines come with cast aluminum, dished-top pistons rated at 8.2: to 8.6:1 compression (stock combustion chamber volumes are listed as 50cc in

Of course custom pistons can be made to your specifications by any of the specialty performance piston manufacturers. This is a forged, dished top, 7:1 compression, blower piston made for a supercharged 229 by Venolia.

If you are using the 4-inch bore heavy-duty block, your choice of performance pistons is nearly limitless. This is a Manley racing piston used in one of the Jenkins test engines.

the 200 and 55cc in the 229 heads).

However, since the 229 V-6 has the same bore, stroke, and pin height as the 305 V-8, TRW's part L2432F, a forged piston for the 305, can be made to work in the V-6. This is a heavy-duty, replacement-type (OEM) piston, with a .110-inch dished top and 1.560-inch pin height, rated at stock compression (8.5:1) in the 305, which should yield approximately 8.75:1 with the stock V-6 heads. The big advantage of this piston is that it is forged instead of cast. Although it is not specifically a high-performance part (it is made primarily for durability under high-temperature operating conditions in the V-8), it looks like the best bet so far in a forged, higher compression piston for the 229. It would also be a wise choice for moderate (5 to 7 pounds boost) turbocharged or supercharged applications, though water injection would still be recommended to insure against piston damage from detonation. This piston can be fitted with TRW's T8337M ring set, which uses a moly top ring.

To fit this piston (or any other V-8 piston) in the 229, you will have to mill .080-inch from the inside of the pin bosses to accept the stock V-6 rods. Although the little end of the V-6 rod is the same width as the V-8 rod, the distance between pin bosses in pistons used for the V-6 is .080-inch wider than it is in V-8 pistons. Apparently this difference is to allow the rod beam to be offset .040-inch in the cylinder bore in the stock V-6. We have received conflicting reports on this issue, and the only true test would be to physically measure the alignment of the rod in the cylinder bore on the engine you are building. Ultimately, this situation would only apply to V-8 pistons used on stock V-6 rods with a stock crank; in applications with a modified crank or V-8 rods, you must determine specific rod location in the cylinder (and consequently in the piston) anyway.

TRW has also recently released a new 12:1 forged pop-up for the 305 V-8, part number L2468F, which will fit the 229 block. However, the slightly smaller V-6 combustion chambers will not only raise the compression about half a point, but might cause some clearance problems with these pistons. Unfortunately, this is a lightweight piston design with a hollow dome, so it cannot be milled down enough to afford streetable compression ratios in the V-6 (although the dome can be massaged somewhat for valve or chamber clearance, plus the V-6 head combustion chambers could be enlarged for piston clearance).

Since the prospects of building a 12.5:1 gasoline engine are severely hampered by the low octane levels of modern standard pump gasolines, and since there is little reason to base an all-out competition engine on the production 229 block, there seems to be limited applicability for this piston

You can even shop for high-performance pistons for the heavy-duty block right at your Chevy dealer. The latest "bowtie" offering is this forged 12.5:1 pop-up.

swap.

If you are working on a 200-inch V-6 and are looking for a forged piston, one good possibility would be TRW's L2401F, listed for the 140-inch Vega four-cylinder. It fits a 3.5-inch bore and has a 1.50-inch pin height. The .060-inch lower pin height is offset by the flat-top design, giving a slight net increase in compression ratio.

The only other alternative, at this time, for performance pistons to fit the production 90-degree V-6 Chevy would be to have something made to your specifications by one of the aftermarket custom piston manufacturers. They should have a good variety of blank sizes available to fit the 200 or 229 blocks.

As we have said, however, most serious Chevy V-6 builders will opt for the heavy-duty, 4-inch bore block to begin with. Combining this block with the big-valve, aluminum Chevy heads allows the use of any of the wide variety of pistons available for the 350 V-8 Chevy, either from the aftermarket

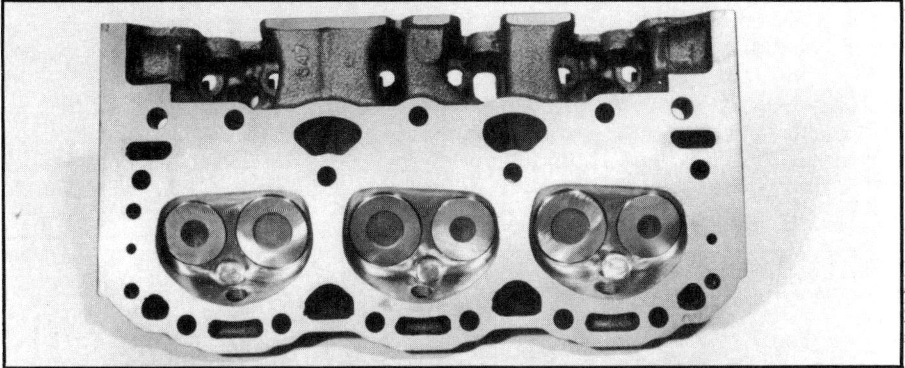

The cast iron Chevy V-6 heads can certainly be reworked for performance, like any Chevy V-8 iron heads. This set has been liberally ported and polished, and fitted with performance valve springs, screw-in studs and guide plates. Valves are stock. Unequal spacing between valve pairs is not apparent without measuring.

or from custom piston makers. TRW alone offers nine different forged pistons for the 350, for applications from street turbocharging through NASCAR racing. Other manufacturers offer similar lines; if you are building a competition or off road V-6, consult them for piston recommendations to fit your specific needs.

Chevrolet also offers a selection of high-performance, off road pistons to fit the 350 V-8, any of which are applicable to the 4-inch bore V-6. Part number 3942541 is a forged 11:1 with pressed pin; 3989048 is an 11:1 with a floating pin. A new Chevy forged, pop-up piston, rated at 12.5:1 and applicable for a variety of racing purposes, has recently been introduced; it will also fit this engine. Part number 14011020 (right side) is for use with 5.7-inch rods; 14011026 is for engines with 6.0-inch rod length.

CYLINDER HEADS

The description of the 90-degree V-6 as "a V-8 with the numbers 3 and 6 cylinders removed" refers specifically to the cylinder head design. Although you might at first think the V-6 heads are identical to V-8's with one cylinder cut off the end, such is not the case. Instead, the V-6 has combustion chamber number 3 sliced out of the left head, and number 6 out of the right

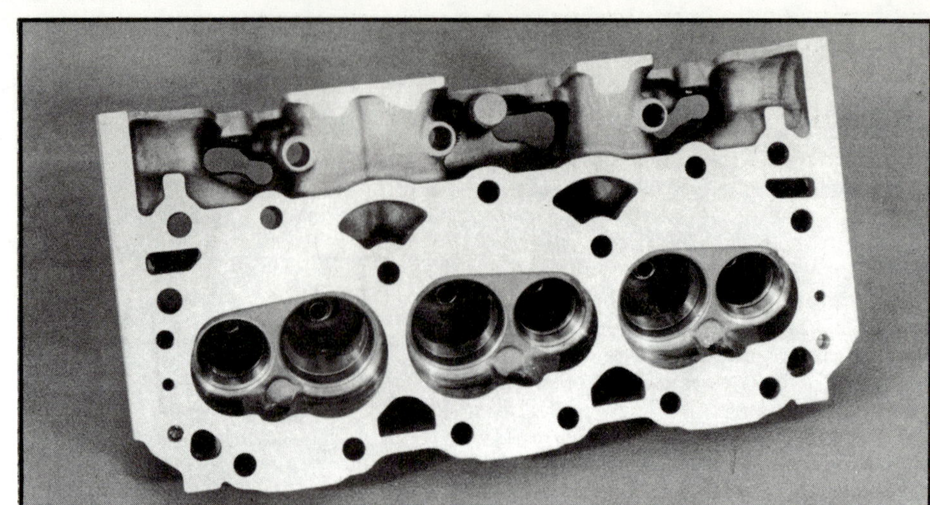

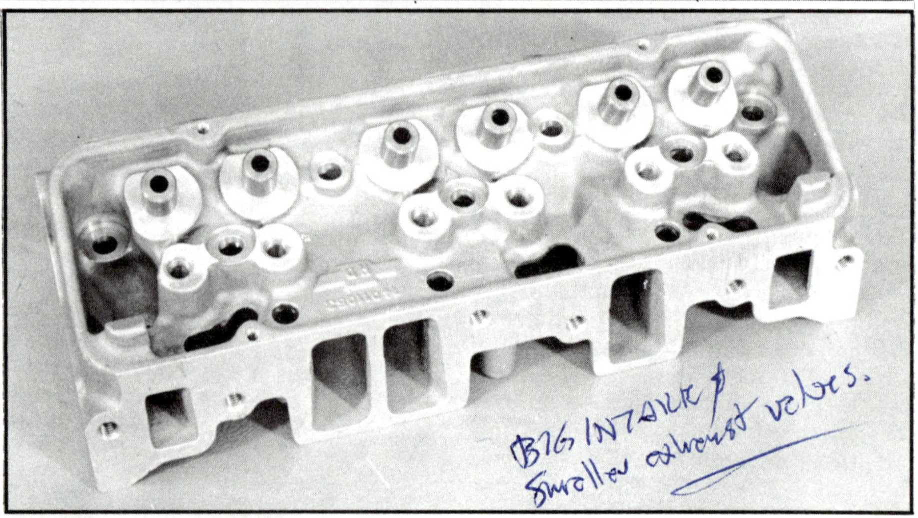

Aluminum Chevy bowtie heads for the V-6 look like this straight out of the box. They're available from your local dealer with large or small valve sizes. This head is for the 2.02/1.60-inch large valves. Note the smaller combustion chamber and elimination of some water passages, compared to stock head.

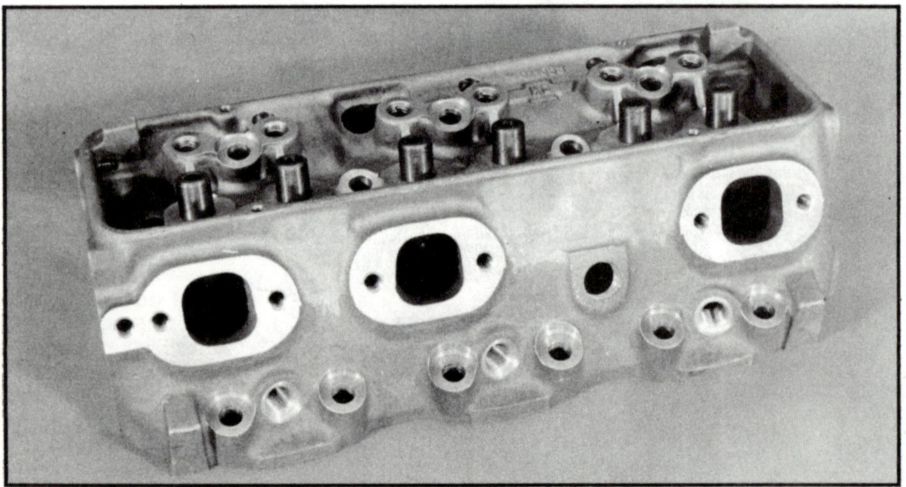

head, yielding a configuration with no two exhaust ports adjacent. The valve arrangement is E-I-E-I-I-E on the left side, E-I-I-E-I-E on the right side (the heads, naturally, will interchange side for side).

Builders comfortable with Chevy smallblock V-8's may find the V-6 head layout difficult to adjust to at first. For instance, spacing between valve pairs is not equal. On the V-8 head, the distance between the middle two pairs of valves (measured from the valve-stems) is less than the distance be-

tween a middle and an end pair. In other words, the space between adjacent intake valve guides and spring seats is greater than that between adjacent exhaust valve guides and springs. This stands to reason since intake valves have larger heads, plus the intake ports must pass between the intake valve guides and pushrod passages in the heads.

Although at first glance the spacing between valve guide, valve spring, rockerarm, and pushrod pairs (as well as lifter bore pairs in the block) ap-

pears equal in the V-6, they are not. As on the V-8, there is a greater distance between the adjacent intake valves (between cylinders 2 and 4 on the right side, and 3 and 5 on the left). However, since the other pair of cylinders does not have adjacent exhaust valves, as on the V-8, but rather an intake opposite an exhaust, the spacing here is slightly wider than that between the middle two valve pairs on a V-8 head.

This may seem like a complicated digression, but it is intended to save you the embarrassment (and expense) of trying to adapt certain V-8 components to the V-6. For instance, you cannot use a smallblock Chevy stud girdle, or rev-kit spring-retainer plate, on the six merely by cutting one end off—the spacing between valve spring pairs is different. Furthermore, while the spacing between rocker studs for any one cylinder is the same on V-8 heads, rocker studs for the middle

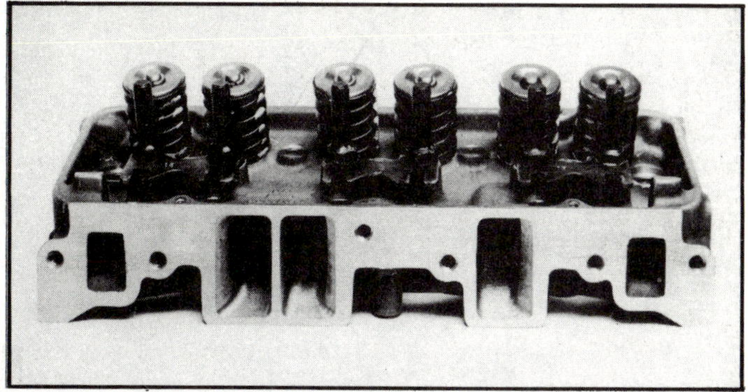

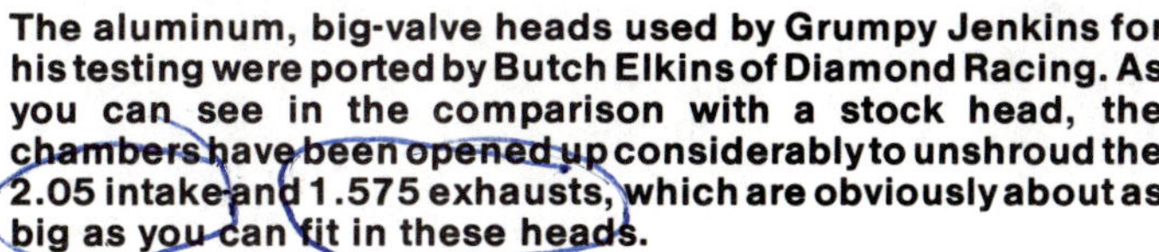

The aluminum, big-valve heads used by Grumpy Jenkins for his testing were ported by Butch Elkins of Diamond Racing. As you can see in the comparison with a stock head, the chambers have been opened up considerably to unshroud the 2.05 intake and 1.575 exhausts, which are obviously about as big as you can fit in these heads.

cylinders on the V-6 are slightly farther apart. On production iron V-6 heads, studs for cylinders 3 and 4 are 1.806 inches apart; end cylinder studs are 1.774 inches apart, as on production V-8 heads. Consequently, V-8-type guideplates can be used on the outer cylinders, but they will have to be cut apart to fit cylinders 3 and 4 on the V-6.

In other respects, the production Chevy V-6 head design is derived directly from the parent 267-305 V-8. One of the exceptionally superior features of the 90-degree Chevy over all other V-6's discussed in this book is the cylinder head bolt layout—this one has 13 attaching bolts per head, while each of the other V-6 designs has only eight. The V-6 uses the same head bolts as the V-8, consequently it could be fitted with a stud kit designed for a smallblock.

On the production heads the port and combustion chamber shape are similar to the 267 and 305 V-8's. The 200 has 50cc chambers with 1.60-inch intake and 1.38-inch exhaust valves. The 229 has approximately 55cc chambers with larger 1.84-inch intakes and 1.5-inch exhausts. The V-6 valves are of the same length and stem diameter as V-8's, so there are lots of possibilities for swapping in larger or stronger valves (limited of course by cylinder diameter). However, we will assume that performance build-ups requiring valve changes or extensive porting will be based on the aluminum heads.

In the summer of 1980 Chevrolet finally introduced their own aluminum cylinder heads for the smallblock V-8...and along with them came similarly designed aluminum heads for the V-6. The six-cylinder heads are available in two versions: part number 14011056 is for "small valves" (1.84-inch intakes, 1.5-inch exhausts); part number 14044802 is for "large valves" (2.02-inch intakes, 1.6-inch exhausts). The small-valve head is designed to bolt on to production 229 blocks, while the big-valve head is designed for the 4-inch bore heavy-duty block. Both feature combustion chambers patterned after the famous "292" V-8 heads (based on casting number 340292), with angled spark plugs. However, the V-6 has a flat quench area under the plug to reduce chamber volume in the small-valve version to 51cc; in the big-valve head the chambers are machined to unshroud the larger valves, increasing volume to approximately 58cc. Both versions are cast with valve guide centerlines spaced .050-inch farther apart to accommodate the larger diameter valves, as on the aluminum V-8 heads. However, on the V-6 the center cylinder rocker studs are still farther apart than the end ones. For pushrod guideplates use Chevy part 14011056 on the middle cylinders (1.840-inch spacing), part 14011051 on the end cylinders (1.8075-inch spacing).

Initial port configuration on the aluminum V-6 heads was patterned on the aluminum V-8 design, though cross-sectional area is slightly smaller to compensate for decreased cylinder volume. These heads feature raised floors in the intake ports, elimination of the water jacket above the intake port roof (it is .200-inch thick), exhaust valve port floors increased .060-inch in thickness, plus increased material (reduction of water-jacket area) around the combustion chambers and valve pockets to allow grinding for larger valves and unshrouding. Another excellent feature of the aluminum V-6 heads, not shared with their V-8 counterparts, is that the exhaust-port flanges have been enlarged considerably to allow drilling for extra manifold-attaching bolts and to provide a better sealing surface for header flanges on turbocharged motors. However, there is no added material behind the enlarged flanges, so holes drilled for bolts (or studs) will have to be sealed since they will enter the water jacket.

The 4th edition of the *Chevrolet Power* manual (released August 1980) gives extensive diagrams and information on porting these heads. However, as of our discussion with Bill Jenkins, in August of 1981, he stated that the aluminum V-6 head castings had been considerably modified. Said Jenkins: "I don't think the original castings are anywhere near as good as they could be. Now modifications have been performed at the foundry to get that stuff fixed so we can do more

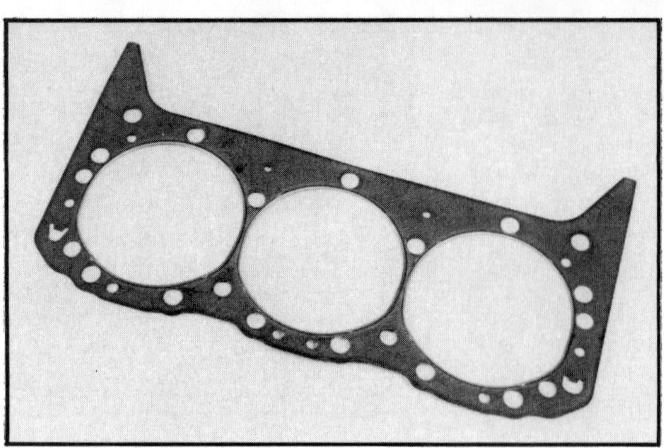

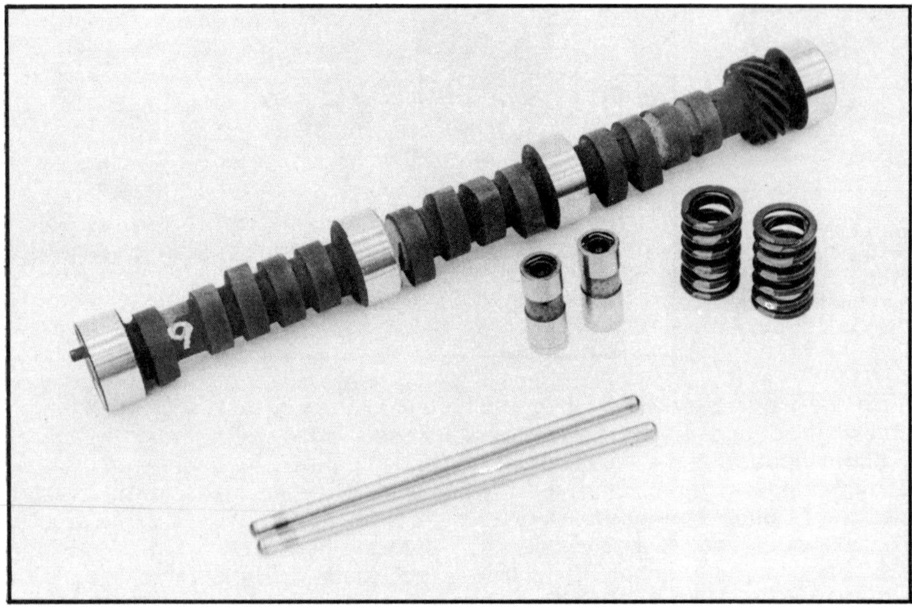

For strong street performance an excellent camshaft choice is Chevy's own "marine" hydraulic grind. This cam and a good variety of high-performance valvetrain components are available from Chevy dealers.

things with them. We just had to go out and hand file the cores to make them look more like we wanted them to be, and have them change the core boxes. We changed the areas underneath the intake port and areas up around the top of the exhaust ports, for grinding—to get you some metal to grind." So the later versions of the aluminum V-6 heads should be better for serious reworking than the early ones.

A further point to consider if you are planning to use the aluminum heads on your V-6 is intake manifold compatibility. The new Chevy cross-ram intake is designed to match the aluminum head intake ports (1.2 inches by 2.2 inches). However, the stock 2-barrel V-6 intake has smaller ports (1.2 inches by 1.98 inches) than the aluminum heads, and will create a port mismatch if installed on them. Whether or not you can enlarge the stock manifold ports to fit the heads is a moot point, since mating this intake to these heads is not very feasible to begin with. No aftermarket manifolds have yet been introduced to match the aluminum heads.

CAMSHAFT AND VALVETRAIN

Camshaft shopping for the 90-degree V-6 could begin right at your Chevy dealer's parts counter, since he lists four different grinds besides the production cam (see the accompanying chart for profile statistics). The more radical Chevy grinds, however, had not actually been released as of our deadline, so make sure these cams are really available before ordering one.

Those of you building mild performance/economy V-6's for street use should take particular note of Chevy's "marine" cam (part 6269733), the primary component responsible for boosting the output of the marine V-6 to 170 horsepower.

Also available at your Chevy dealer are compatible valvetrain support pieces. For street performance use valve spring and damper set 3927142; for moderate racing use the 330585 dual-springs with the 330586 aluminum retainers; for big-lift cams and roller grinds use dual spring (plus damper) set 366282 with the 366254 titanium retainers. To maintain correct rockerarm geometry with high-lift cams (over .450-inch, which requires a smaller cam base-circle diameter), use Chevy 366277 heavy-duty pushrods, which are .100-inch longer than stock and have .075-inch wall thickness. With milder cams you can use the Chevy "blue-stripe" pushrods, which are standard length with .060-inch wall thickness. Install these pushrods with the insert tip end up. For converting production heads to screw-in rocker-arm studs, or for the aluminum heads (which come drilled and tapped), Chevy offers part 3973416 (3/8-inch) or part 3921912 (7/16-inch) heavy-duty studs. As noted, the aluminum heads require special pushrod guide-plates; part 14011051 for cylinders 1, 2, 5, and 6 and part 14011057 for cylinders 3 and 4.

When installing any high-performance valve springs on the production V-6 heads, be sure to check the installed setup for correct spring height and adequate clearance between coils (at least .040-inch total) in the fully open valve position to guard against coil bind. The stock heads use "rotator" retainers on the exhaust valves only, which reduce spring height about .030-inch. On early 200 heads, the exhaust valve seat is lowered this amount to allow the use of same size springs on both intake and

CHEVROLET 90° V-6 CAMSHAFTS

Part Number	Description	Duration (total)	Lift (net)
14014447	Production 299	300°int/310°exh	.357int/.390exh
6269733	Marine, hydraulic	320°int/334°exh	.410int/.410exh
14011097	Off Road, hydraulic	342°int/342°exh	.447int/.447exh
14011098	Hi-Perf., solid	Not available	Not available
14011099	Racing, mushroom	Probably will not be released	Probably will not be released

Note: All measurements taken at zero lash.

exhaust. On later models, the exhaust valve springs are shorter than the intakes. For a performance application, remove the rotator, using a standard (or high-performance) retainer on the exhaust valve as well as the intake. If your heads have lowered exhaust spring seats, use shims to make them equal to the intake side. *Never* install performance (or stock, early V-8) valve springs with rotators, since this can lead to coil bind, especially with higher lift cams.

Besides the stock hydraulic lifters, Chevy lists only one other lifter for the V-6, a piddle-valve, mushroom tappet (part 14011080) for use *only* with a cam designed for mushroom lifters—and such a combination would have limited applicability, such as NASCAR racing where rollers aren't allowed. Although any V-8-type lifters will fit in the V-6 block, the enlarged right-hand lifter oil gallery in all GM V-6 blocks causes specific concern in lifter selection. An edge-orifice solid lifter should not be used in these blocks. Further, Chevy recommends against using the big-block, piddle-valve solid lifter, suggesting instead the Sealed Power part AT-992 lifter for reduced pushrod seat leakage. For hydraulic-lifter cams the TRW VL66RH "high-rev" lifter is compatible with the V-6 oil system.

The enlarged oil gallery requires the use of special-design roller lifters (as pointed out in other sections of this book). Crane currently offers rollers with a "hooded nose" for use in the V-6. These should work fine with moderate lift roller cams, but with extreme grinds, you will have to do some close checking of the assembly to guard against pressure leakage from the all-important left hand gallery at the highest and lowest point of lifter travel.

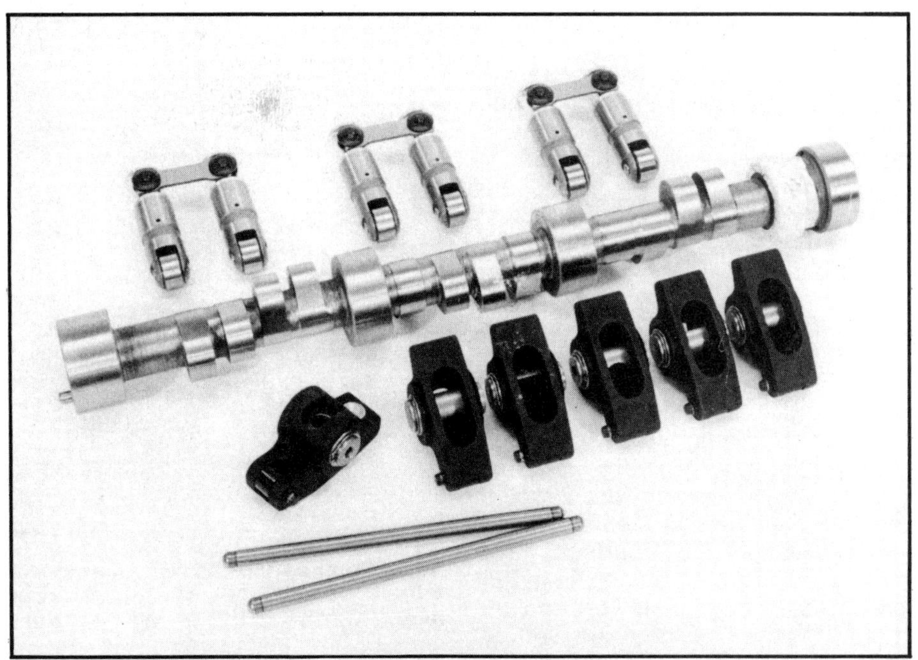

This Iskenderian roller was one of the first billets available for the Chevy 90-degree. Shown with it are typical Isky smallblock Chevy support pieces such as aluminum roller rockers and one-piece 5/16-inch pushrods. To use the Chevy roller lifters shown, however, you would have to convert a heavy-duty block to V-8-type oiling.

As described by Bill Jenkins: "The base circle of the cam can be critical on this. The lifter will drop down in the lifter bore quite a bit if the base-circle diameter has been reduced to gain a lot of lift on a regrind. There is a point where it might get critical if the lifter is dropping down too far. If there is any small error in the machining of the block or the casting at the top of the lifter bosses, if it gets too short or something like that, it will spit oil out of there like crazy. Now that is definitely critical. So you have to make sure there is no oil blowout on all six holes on the left side. None of the blocks I've seen so far leak out there, but they get very, very close. In some cases, if the lifter dropped .035- to .040-inch more, they would leak."

Billets for aftermarket Chevy V-6 cams have recently been introduced, which should reduce this problem somewhat (although the base circle will still have to be reduced slightly on super high-lift grinds, since maximum cam-lobe height is limited to cam bearing journal diameter). And, since the V-6 accepts most other V-8 valve-train pieces (with exceptions already noted), the aftermarket can provide a vast array of high-performance cams, lifters, pushrods, valve springs, retainers, rockerarms, studs, valve

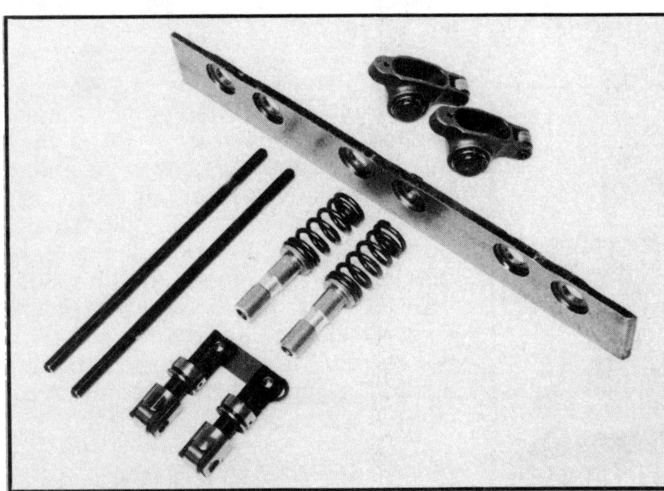

Valvetrain pieces used by Grumpy Jenkins for his Chevy V-6 testing included Crane rollers, Crower steel rockers, Chevy heavy-duty pushrods, and a rev kit with hand-cut steel retainer plates (since none were available at the time and cut down V-8 units will not fit the V-6).

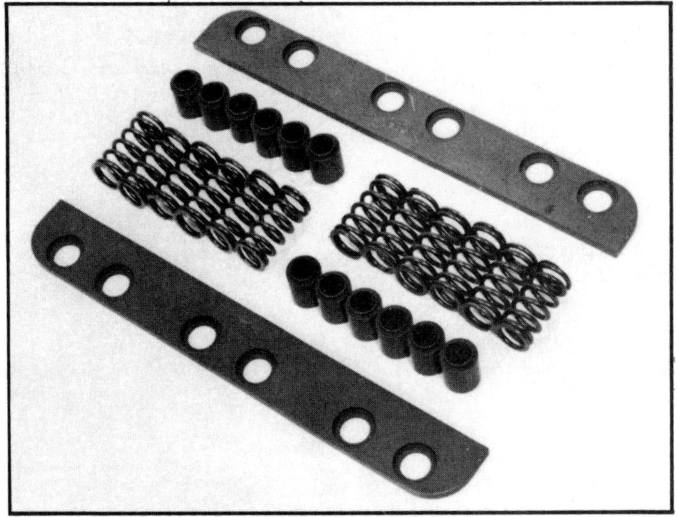

Now a complete rev kit is available for 90-degree Chevies from Baker Engineering. The retaining plates are made from blue anodized machined aluminum.

guides, seals, valves, thrust bumpers, and so on.

Valve sizes in the 229 are limited to a maximum of 1.88 inches intake and 1.55 inches exhaust by cylinder-bore diameter, which limits availability of ready-made performance valves, but larger valves could be cut down to fit the small-bore engine. Otherwise, the V-6 will accept just about anything that state-of-the-art Chevy smallblock valvetrain technology has to offer.

Incidentally, the V-6 was originally fitted at the factory with excellent heavy-duty, aluminum-faced, tri-metal cam bearings that are superior in quality to those used in production V-8's. For a while the hot tip for V-8 racers was to install V-6 cam bearings; now the production V-8 bearings have been upgraded to the same quality. So there is no need to install better cam bearings in the V-6. If they are worn or damaged, TRW's replacement part number for the V-6 is SH1089S.

As far as camshaft profile and lobe spacing is concerned, the firing order of the V-6 (which produces an evenly spaced firing sequence in each cylinder bank, unlike the V-8) will take some further thinking and experimentation to find out what really works the best in this motor. For his V-6 test program, Jenkins used a Crane 270-417-2S6 track roller grind (270 degrees duration, .417-inch lift measured at .050-inch lifter rise, with 106-degree lobe centers) and saw indications that the engine could use even more duration (276 degrees). When asked what present V-8 technology would not translate directly to the V-6, he had this to say: "I don't really have enough experience to tell you that from a tuning standpoint. I expect that with the even firing down each bank there will be some change of cam-center requirements and header con-

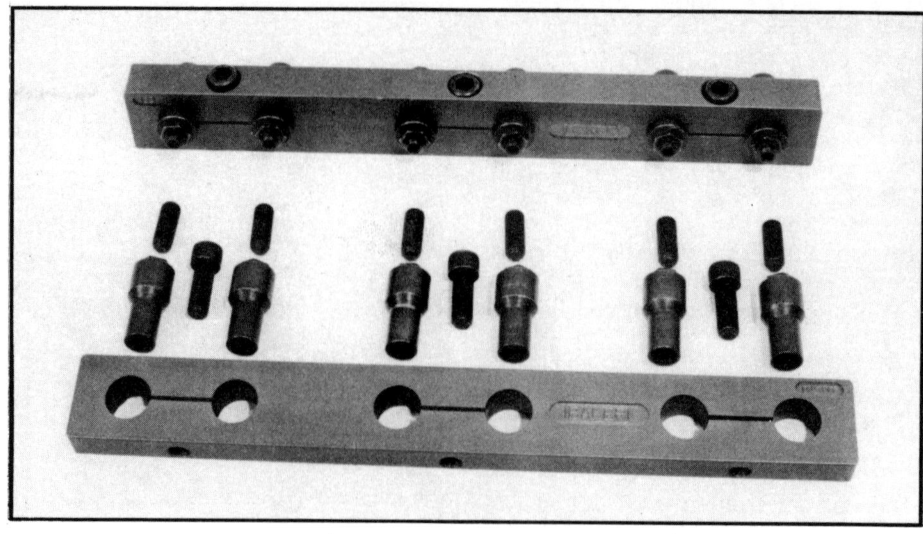

Baker Engineering has also recently introduced rocker stud girdle kits for the Chevy V-6 in sizes to fit 7/16- or 3/8-inch studs on Chevy bowtie or new Brodix V-6 aluminum heads. Girdles to fit iron V-6 heads are made by Ridgeway.

The 90-degree V-6 can use any timing gears made for the smallblock V-8, but a good choice for street or racing is the Chevy double-roller chain and gear set (part numbers 3735411 chain, 3735412 cam gear, 3735413 crank gear) used on Chevy pickups.

The only aftermarket intake manifold available for the 90-degree Chevy is this Street Dominator 4-barrel just released by Holley. It is a street/recreation type of intake designed for use with a small spread-bore such as Holley's model 4360. The ports match the production cast iron heads.

ditions, and you may be able to use more intake size, intake port runner, and so on in that motor—all of the things that generally go with the even firing pattern. Now when you get to the cross-ram manifold, which increases port runner length, your cam-center requirements change quite a bit. But even without that, I'd say that your existing knowledge about cam and header stuff is not going to directly apply. It's going to take some more work to see what's going to apply."

INTAKE SYSTEMS

Like the other V-6's discussed here, the 90° Chevy uses a sequential firing order and is best suited with a single-plane, dual-log intake. The stock manifold is aluminum and mounts a Rochester #210 2-barrel carb. On 200-inch engines, the manifold has a center divider, with each carb venturi feeding one cylinder head in strict dual-log fashion. The 229 manifold has a completely open chamber under the carburetor pad, resulting in a "360 degree," single-plane design. Although both manifolds look the same on the outside, you can check for the divider with your finger through the carburetor flange holes (with the carb removed). This is one good way to distinguish between a 200 and a 229 in the junkyard. All 1982 and later passenger cars using the 90-degree V-6 will of course be fitted with a "tamper-proof" computer control system, as discussed in the 60-degree engine section.

Holley has just released a replacement four-barrel intake manifold for the 90-degree Chevy. It is an RV-type intake similar to the Street Dominator for the Buick V-6, designed to accept the Holley 450cfm model 4360 or other spreadbore, 4-barrel carburetors. This manifold is cast to match the stock production head ports; it could be opened with a grinder to match unported aluminum heads, but would require welding on the tops of the flanges for anything larger.

If Chevy does in fact produce a 262-inch version of the V-6 for 1983 vans, a 4-barrel intake is slated to come with it, which means you should be able to readily purchase one at your Chevy dealer or at wrecking yards in the future. If the project follows current plans, this intake will be a hi-rise design, which should improve fuel distribution in the V-6, and it will be fitted with a Rochester Quadrajet carburetor. The van body for which this engine is intended allows room for a hi-rise induction system; it would probably cause hood clearance problems if

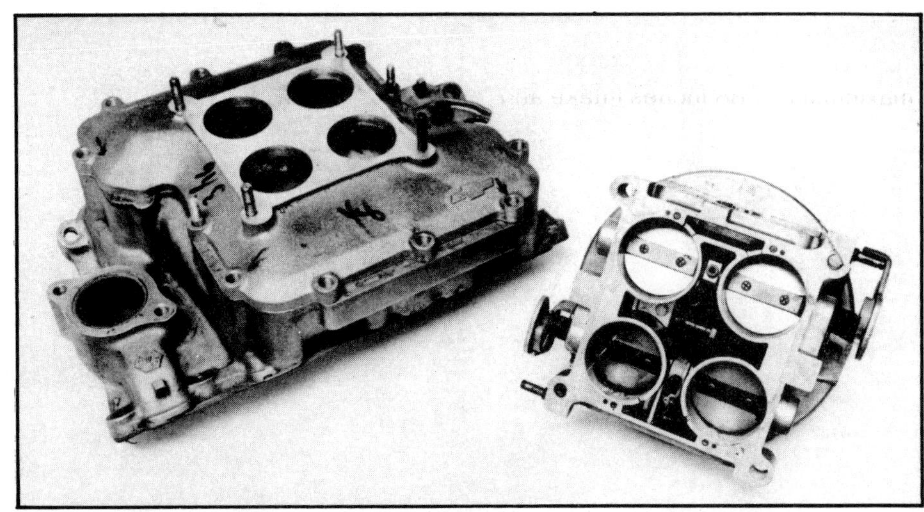

The only racing intake for the 90-degree is offered by Chevy for offroad use. It resembles the fabled "rat roaster" cross-ram, and is obviously aimed at track competition (as opposed to street or drag racing). It comes with a single 4-barrel top, this one having been bored out to accept a Holley Dominator, and produces a strong, wide torque band.

installed on a late-model passenger car. Of course all of this is conjecture until the parts are actually put into production by Chevrolet.

The only competition manifold currently available for the Chevy is the factory two-piece, box-plenum item, designed much like the old Z-28 cross-ram, but with a single 4-barrel top. The complete manifold is composed of base section 14044804, bottom oil splash shield 14044805, plenum cover plate 14044804, and 4-barrel top 14044803. This is an all-out racing manifold, designed primarily for high-rpm use on oval tracks (i.e., NASCAR),

The Chevy cross-ram uses a bolt-on center divider, shown removed in left photo. To achieve impressive dyno figures, Jenkins and Butch Elkins spent considerable time whittling away at the stock casting. Note how much the runners have been opened up on the modified version (right photo).

although it has shown an amazingly wide and flat torque curve in tests, making it very suitable for road courses. It is cast to match the big ports in the aluminum heads, it has no heat passages, and it would certainly be incompatible with conventional street use.

For his V-6 tests, Bill Jenkins used the Chevy cross-ram manifold bottom (extensively reworked, as shown in photos), combined with a hand-cut, .50-inch spacer ring and a .50-inch, flat aluminum top cover plate milled .300-inch on the bottom side. This custom top plate allowed Jenkins to use a pair of 2-barrel carbs, rather than a centrally-mounted 4-barrel (as required by many racing classes), for more optimum fuel distribution with this manifold design. Plus, he drilled the top plate to try different carb locations, finding that a central spacing, with carbs directly across from each other, worked the best. However, using the

Jenkins felt a pair of 2-barrels would work much better than a single carb on this manifold design, and built his own top to test this unusual arrangement. The only problem was finding 2-barrels big enough to feed the 437 horsepower six.

largest 2-barrels he could find (1-9/16-inch venturi Holleys—the end carbs from a Mopar 440 Six-Pack), Jenkins felt the 265-inch, 430-plus horsepower V-6 could still stand more carburetion, and he was therefore considering going back to the single 4-barrel top with a 4500-series Holley Dominator.

Jenkin's tests were conducted specifically for oval or road racing results, "looking for as much width to the power band as possible." The tests netted a broad, flat torque curve which hovered right at 350 ft-lbs from 4500 to 6500 rpm. As Jenkins said, "That manifold gives an unusually wide torque range. The peak torque curve we got is *outstanding* relative to anything. You'll seldom see a flat-top torque curve like that. We attribute most of this to the manifold."

However, Jenkins also noted that this particular manifold was a limiting factor as far as peak horsepower was concerned. "It's specifically a round-track manifold, and that's all." For drag racing, where horsepower is far more important than torque, and where class rules allow more exotic induction systems, he states that "a tunnel-ram is probably the only way to go." Since no tunnel-ram intakes are available for the V-6, you would have to adapt one from a smallblock V-8, but that shouldn't be overly difficult since the intake port spacing on the V-6 is nearly the same as on the V-8. Says Jenkins, "Cut up an old Pro Ram II or something and make one—that would work out as well

as anything." The same should hold true for direct-port fuel injection systems. A variety of injection setups are available for the smallblock Chevy, any of which could be cut and adapted fairly easily to the six. Fuel pump drives, front covers, and other ancillary pieces will adapt directly to the V-6.

IGNITION

Ignition timing is a concomitant problem to the 90-degree V-6 crankshaft dilemma. The 18-degree pin splay of the Chevy V-6 crankshaft design falls between a fully odd-fire, shared-pin configuration and the Buick's 30-degree split, even-firing crank. But the fact remains that the Chevy is not an even-firing motor, and this can lead to ignition problems under extreme racing circumstances. Going to 30-degree offset pins to produce an even firing pattern would weaken the crank if rod-journal diameter were not also increased. Going to V-8-type common pins would solve the crank, rod, piston and bearing misalignment problems, but would only worsen the ignition situation.

Bill Jenkins has several pertinent comments about the ignition situation for the Chevy V-6, so we will excerpt from the interview:

Question: If you're weighing the two, crankshaft strength over ignition stability, do you see a point at which one would take precedence over the other?

Jenkins: Well, for drag racing the ignition is pretty important because you like to run a lot of plug gap and you *need* an MSD or something like that to fire a wide gap. If they could produce any kind of sorting device that could fire the thing in uneven intervals—do it

For drag racing a tunnel ram of some sort would work best on the V-6. As Jenkins suggests, an Edelbrock Pro Ram or Victor Ram for a smallblock V-8 can be sectioned to fit the V-6 and welded back together.

While Maropulos used a single 4-barrel on his sectioned tunnel ram, Rick Voeglin mounted one Holley 4-barrel and one 2-barrel, as shown, to feed the six runners directly. Note slight angle of carbs to align with ports.

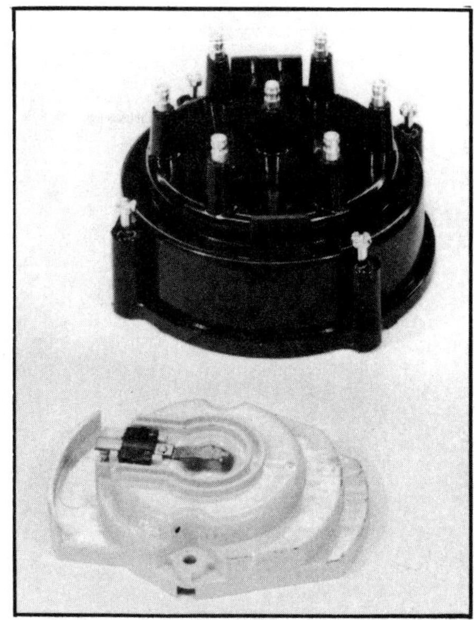

Jenkins used a Chevy inline-six cap with external coil on the V-6 HEI body. Wide rotor tip aligns odd firing with even terminals.

reliably and do it inexpensively—I don't know. But if you want to get decent ignition in the motor, and you want a short-stroke motor, then you just build an even-firing crankshaft.

Question: What do you see as the best ignition possibilities, now, using the stock-type crank?

Jenkins: The only thing you can use at this point in the game is the Delco HEI. On our test motor we used the HEI with short top, anchored down the reluctor so it didn't have any vacuum advance, and modified the centrifugal advance slightly.

Question: Did you have any problem with the HEI at all?

Jenkins: Well, it won't fire a lot of plug gap. It's necessary to run a gap of not over .030-inch. But with the small gap, it'll run the motor fine at 8000rpm.

There's not really much more to say on the subject, unless MSD, Accel, or one of the other aftermarket ignition specialists comes out with a system designed especially to cope with an odd-firing engine. The only other possibility, especially for a big-inch drag motor, would be to build your own crank-trigger system using two pick-ups and two separate ignitions, one to fire each bank of three cylinders.

However, it should be stressed that ignition problems occur in an odd-fire engine only under very extreme circumstances. As Jenkins pointed out, the Delco HEI will work very well, if set up properly, to limits that very few V-6 drivers will ever see.

HEADERS

As far as we can ascertain, no one is making bolt-on headers for this V-6. (When looking through header application charts in catalogs, be sure not to mistake the "corporate" V-6—that is, the 231-inch Buick—for the Chevy 229, both of which are sometimes referred to as 3.8-liter engines. Both have been installed in the same body styles in some instances, but headers are available only for the Buick so far.)

So if you want a set of headers for your Chevy V-6, you will have to have them fabricated. Although they are usually expensive, custom header fabricators can be found in most major cities. If you are going to install the headers on a stock-block motor in a street vehicle, you will want relatively small primary tubing (1.5- to 1.75-inch in diameter) relatively long in length (30 to 36 inches) with something like a 3-inch (o.d.) collector. However, in

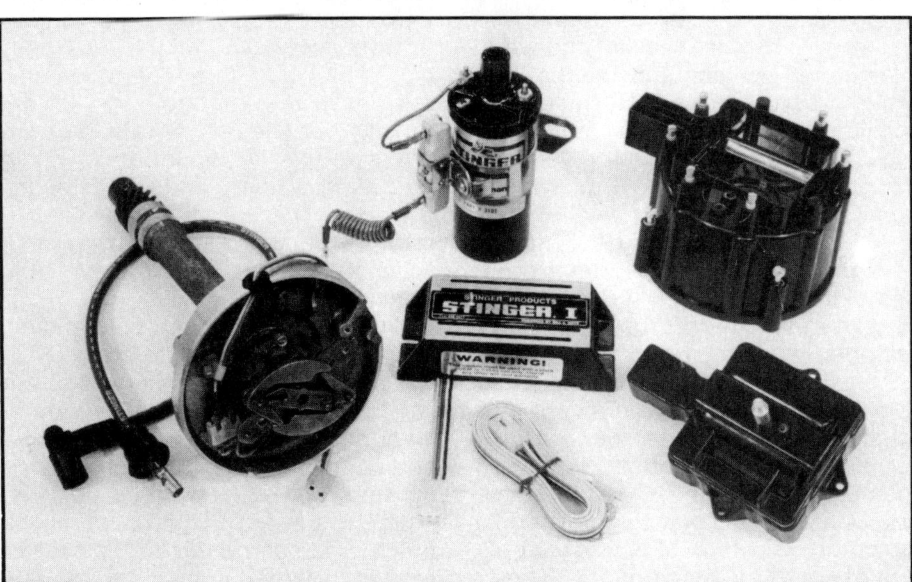

Most aftermarket ignition companies, such as Stinger, now offer a variety of street performance and racing conversion systems for the GM HEI distributor body.

53

Do not forget that the 90-degree Chevy is an odd-fire motor. Even though distributors from inline Chevys, such as this Vertex magneto, will drop in and run the engine—they won't run it correctly.

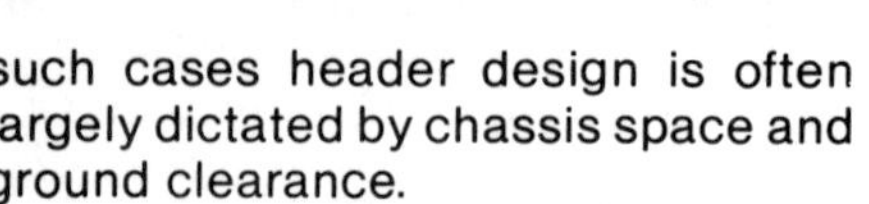

Moroso offers this 6-quart extended pan for the V-6. The 90-degree will also accept small-block V-8 performance oil pumps.

Another recent Chevy bowtie performance part that will fit the 90-degree motor is this aluminum water pump, which features extra bracing around the shaft.

such cases header design is often largely dictated by chassis space and ground clearance.

On his comparatively large displacement test engine, Grumpy Jenkins used 1.75-inch by 26-inch header pipes funneled into 3-inch by 44-inch open tail pipes, as would be used on a track car. "We messed around with the headers a little bit," said Jenkins. "Ran it without the tailpipes and found a couple of peaks in there that weren't there with the tail pipes, but the torque curve wasn't anywhere as nice for a round-track car." When asked about "phasing" the headers on an odd-fire motor, Jenkins replied: "You're talking about two different things here. The even firing pattern is always the same regardless of the crankpin situation because the three cylinders fire in even spacing on each side. On a V-8 they don't. And that has nothing to do with the crank phasing. That has to do with the phasing from one side of the block to the other."

Since all of the V-6's discussed in this book use the same basic firing order, they all have the same "phasing" and the same header technology should apply to all of them. Smokey Yunick had originally designed a "uni-pipe" exhaust system for the Buick V-6, with all six pipes gathered into one collector behind the engine (for possible Indy car application); later he developed extremely short "W"-shaped headers for this engine, with a separate header for each side. Asked for

his comments on these designs, Jenkins replied, "I'd have to be convinced the uni-pipe thing works. It's hard to put more than four cylinders into one pipe and get the collector effect to work for you. I have never tried the short header pipe system, so I have no opinion on it."

BOLT-ON PARTS

Since the 90-degree Chevy V-6 will accept several smallblock V-8 components, a vast array of pre-existing high-performance parts is available for this motor, and it would be nearly impossible to list all of them here. However, the basic similarity between unique V-6 components and their V-8 counterparts has induced some companies to tool up for a few new parts specifically for the V-6. Moroso (Guilford, Connecticut) lists a 6-quart deep sump oil pan for the Chevy V-6, part 2033. With a total depth of 8.25 inches, it might cause ground clearance problems in some applications, and it requires extended pickup part 2417 or oil pump spacer kit part 2310 with a stock or aftermarket pump (such as Moroso's 2210 racing oil pump). The stock V-6 pan holds only four quarts of oil, and the use of a high-volume oil pump is not recommended with this pan, since such a pump could literally suck the oil pan dry. For street applications with the stock pan, TRW recommends oil pump 50065, which is a Z-28 pump delivering 55-60psi pressure (or use the Chevy 3848907 high-

pressure pump). Moroso also lists an aluminum water pump, part 6350, with a reinforced housing and low-drag impeller, plus Chevrolet has recently introduced their own aluminum water pump, part 14011012.

Custom valve covers for the Chevy V-6 are plentiful. Moroso offers extra tall covers (to clear rocker girdles) in gold-anodized, stamped aluminum (part 6825) or chrome-plated steel (part 6826). Chevrolet also plans to introduce tall, chromed-steel "Chevy Power" valve covers for the V-6. Holley has finned, cast-aluminum valve covers similar in design to those for the Buick V-6, while Blower Drive Service offers tall cast aluminum covers made by cutting, welding, and repolishing V-8 covers. The V-6 uses RTV sealant rather than gaskets, but good cork valve cover gaskets are now available from Mr. Gasket.

In the clutch, flywheel, bellhousing, and transmission department you can choose from anything that will fit the smallblock V-8, with the exception of large-diameter, big-block flywheels and matching starters. To use such a flywheel (and larger clutch) on his test motor, Grumpy Jenkins machined a notch in the starter mounting area of the block, added a small bolt-on tab, and adapted a big-block starter to the V-6, thus allowing the use of the larger flywheel.

SUPERCHARGING, TURBOS, AND NITROUS-OXIDE

So far no special kits have been developed for bolting a turbocharger or nitrous-oxide injection onto the Chevy V-6, but either system can be readily and easily adapted to this engine with components currently available. For under-the-carb nitrous injection a special carburetor plate would have to be fabricated—most of the nitrous suppliers could have this done. For a street turbocharged application you might want to consider a higher volume oil pump, such as the TRW part 50135, since extra engine oil is circulated through the turbo.

The real news in "abnormal aspiration" for the 90-degree Chevy is from Blower Drive Service (Whittier, California), who built their own version of a 3-71 GMC supercharged, 200-inch Chevy by adapting a blower mount to the stock intake manifold. Response was significant, so BDS has recently designed and produced an all-new, cast-aluminum blower manifold to fit the Chevy V-6. It accepts a 4-71 GMC blower, and has a standard water outlet and a pop-off valve. Port size matches the production heads, but the manifold can be opened up to match the aluminum head ports. BDS also offers a full line of prepared superchargers and drive components to make a complete package for the V-6.

Blower Drive Service offers a complete line of reworked GMC blowers, engineered drive sets and installation kits for adapting street or competition superchargers to Chevy, Buick or Ford V-6s.

Heart of the BDS kit for the 90-degree Chevy is this cast aluminum 4-71 manifold. It incorporates a stock-type water neck at the front and a pop-off valve at the rear.

This early version of a 4-71 blown Chevy V-6 was built by Blower Drive Service when the 200-incher was first introduced. It used a modified stock intake manifold. The tall cast aluminum valve covers, made by cutting down V-8 units, are available from BDS.

CHEVROLET 90° V-6 PRODUCTION SPECIFICATIONS

	200 cubic inch engine	229 cubic inch engine
Displacement	3.28L (200ci)	3.75L (229ci)
Bore	3.50-inches (89mm)	3.74-inches (95mm)
Stroke	3.48-inches (88.4mm)	3.48-inches (88.4mm)
Years produced	1978-1979	1980 and later
Firing order	1-6-5-4-3-2	1-6-5-4-3-2
Cylinder numbering	1,3,5 left bank 2,4,6 right bank	1,3,5 left bank 2,4,6 right bank
Compression ratio	8.2:1	8.6:1
Main journal diameter	2.45-inches (62.23mm)	2.45-inches (62.23mm)
Rod journal diameter	2.10-inches (53.34mm)	2.1-inches (53.34mm)
Rod journal offset (split)	18°	18°
Connecting rod length	5.70-inches (144.78mm)	5.70-inches (144.78mm)
Valve head diameter	Int: 1.60-inch (40.65mm) Exh: 1.38-inch (35.05mm)	Int: 1.84-inch (46.74mm) Exh: 1.50-inch (38.10mm)
Rockerarm ratio	1.5:1	1.5:1
Valve clearance	Zero	Zero
Main bearing clearance	.0020-.0035-inch	.0020-.0035-inch
Rod bearing clearance	.0010-.0025-inch	.0010-.0025-inch
Piston-to-bore clearance	.0007-.0042-inch	.0007-.0042-inch
Piston ring gap	.010-.020-inch	.010-.020-inch
	Bolt torque (pound feet)	Bolt torque (pound feet)
Cylinder head	65	65
Main caps	75	75
Rod caps	45	45
Intake manifold	30	30
Flywheel	60	60

CHAPTER 4

BUICK V-6

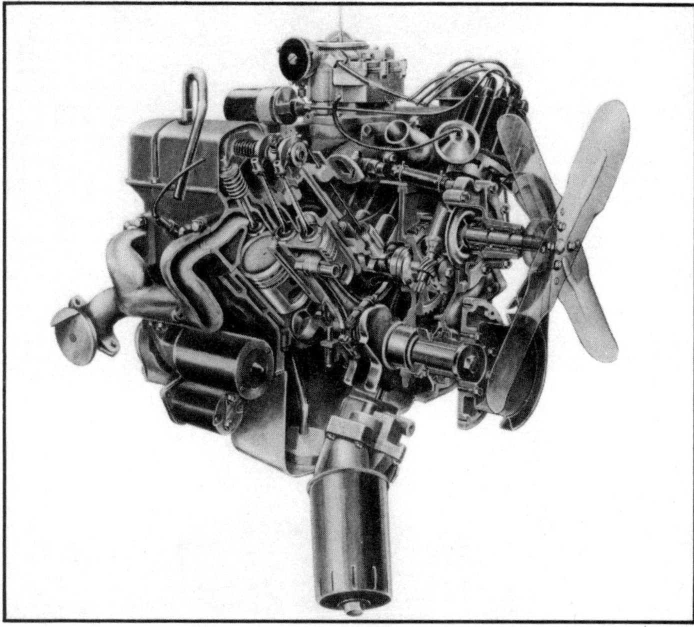

BIRTH OF THE BUICK

The Buick V-6 has been around, on and off, for over twenty years. Unlike other engines discussed in this book, which hitherto have received little press and scant in-depth performance analysis, the Buick V-6 has been given extensive coverage, especially since its re-adoption by General Motors in 1975 and avid promotion by Buick as the performance engine of the future. Consequently, we will not attempt to present here an exhaustive study of all aspects of possible Buick V-6 racing applications, but rather a comprehensive look at current aftermarket parts availability for this motor, an update on factory components, and a thorough discussion of building techniques and tips that pertain primarily to street, off-road, and recreational racing purposes.

The short story of the birth of the Buick 90-degree V-6 is surprising and it reinforces several of the points made in our introduction concerning V-6 design. Early in the 1960's Detroit was caught up in its first small-car era, prompted by a short-lived national economic slump. Not only were the manufacturers building compacts, but they even offered a *range* of the smaller cars, including mid-luxury models like the Buick Special/Skylark and the Olds F-85. These two compacts were designed around the all-aluminum 215 cubic inch V-8—a light, powerful, fuel-efficient engine perfect for such a vehicle, but one which was very expensive to produce. When the car-buying market deteriorated further, Buick decided that they would have to invent a cheaper powerplant for their Special, and they'd have to do

When Buick introduced America's first passenger car V-6 in late 1961, it had a Rochester 2-barrel, points-and-coil ignition, moderately dished pistons and cast-aluminum non-adjustable rockerarms. The odd-fire crank had paired rod pins (like a V-8). Note the hefty triangulated generator bracket—a necessity to keep it from breaking due to engine vibration.

it in a big hurry. Assuming that the American public wasn't ready for four cylinders yet, they decided on a six. The problem was that the car was designed for a V-8, and a straight six wouldn't fit. GM had been well aware of the V-6 configuration at least since the late 40's, but they were leery of vibrations in such an engine without a counterbalance shaft (and building a balance-shaft V-6 could be as expensive as building the aluminum V-8). But in 1960 GM had introduced a big 90-degree V-6 for pickups and other

When Buick re-released the V-6 in 1975 the only substantial changes from the original 1961 design were stamped rockerarms, slightly larger and deeper dished pistons, a breakerless HEI distributor and an alternator.

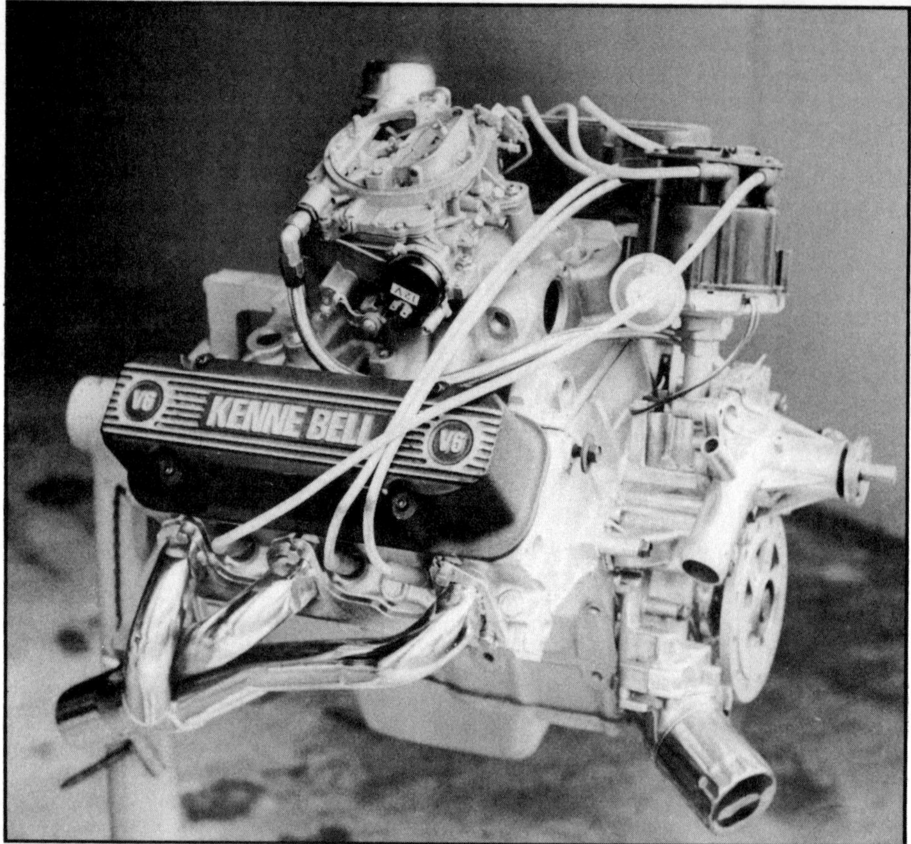

A V-6 provides sluggish performance in a heavy car, but so do all "modern" over-regulated, low-compression engines. With the addition of a few traditional bolt-on performance parts—headers, cam, carb, manifold, and ignition...all of which are readily available for the Buick—doubling the stock 110 horsepower is not out of the question.

The Buick V-6 is a very practical and popular powerplant for early street rods, particularly those with small engine compartments—like Model-A's and Model-T's. With a few aftermarket components the V-6 not only looks like a hot rod motor, but in a light car it acts like one—and offers amazing mileage at the same time.

trucks and it had met with success, so Buick decided to take the plunge.

From initial conception in March of 1961 to the new car introductions in the fall of that year, the gestation period for the new Buick V-6 was only six months. Given normal development time for new engines, that probably still stands as a Detroit record. Of course, the V-6 wasn't really an entirely new engine. Buick simply took their V-8, chopped off two cylinders, and cast it in iron instead of aluminum.

That's oversimplifying a bit, but it demonstrates aptly how and why the V-6 was born, and how the first American version came to be cast in a 90-degree Vee configuration. The '61 Buick V-6 was significantly less expensive to build than the aluminum V-8, it was only 50 pounds heavier overall (362 pounds total), and it could use nearly all of the existing V-8 assembly components, with the notable exception of the crank, cam, rods, pistons, heads and manifolds. Bore and stroke were increased from 3.5 inches by 2.8 inches in the V-8 to 3.625 inches by 3.20 inches in the initial V-6, giving a displacement of 198 cubic inches.

Although individual cylinder volume was substantially increased in the V-6, the heads retained the same valve and port sizes as the V-8. This was one drawback of the rapid development program which limited performance potential on these early engines and affected the V-6 for nearly 20 years. Other vestiges of the parent V-8 still haunt the V-6 to this day (remember, the current engine is basically a 25-year-old design): the extended-skirt block (fine for an alumuinum motor, but dead weight on the six), a rockershaft system prone to flexing and cracking, and an aluminum front housing which mounts the distributor, oil filter, and water, oil, and fuel pumps in one unit. The front cover design, though it provides definite advantages in terms of engine assembly and maintenance, adds extra length to this V-6, as well as saddles it with a convoluted and controversial oiling system. We will discuss these characteristics and explore performance-oriented compensations in the following sections.

Besides eliminating two cylinders from the V-8, the Buick designers also had to redesign the crankshaft for 120-degree spacing between each of the three throws (they did not consider the split-throw configuration at that time), and tailor a camshaft with appropriate timing sequences. Given the three-throw crank, they could have chosen between two basic firing patterns for the V-6. One, with a firing order of 1, 6, 2,

4, 3, 5 would produce 150°, 120°, 120°, 90°, 120°, 120° crankshaft intervals between firings. It would require a true dual-plane intake manifold, with one side feeding cylinders 1, 2, and 3; the other feeding cylinders 4, 5, and 6, and it would produce uneven exhaust pulses in each manifold. The other choice, a 1, 6, 5, 4, 3, 2 firing order, would give a 150°, 90°, 150°, 90°, 150°, 90° sequence, producing even firing pulses in each bank of three cylinders.

The latter pattern was chosen for the initial odd-fire motor because (1) overall torque fluctuation throughout the cycle was smoother than the version with three different fluctuations, (2) the even spacing of the exhaust pulses in each manifold was preferable, and (3) even intake spacings in each head allowed the use of a simpler single-plane intake manifold, divided into two sections, with each half feeding one cylinder bank from one venturi of a 2-barrel carburetor. Although the crank pins were later split to produce a fully even-fire V-6 motor, the 1, 6, 5, 4, 3, 2 firing order has remained the same (and is the same for all V-6's discussed in this book) for the second and third reasons cited above.

The new V-6 was both an engineering breakthrough and an economic achievement for Buick in model year 1962. In an introductory article in HOT ROD magazine (December 1961), Roger Huntington saw great things in the immediate future for this motor: "...And who knows what might happen

A V-6 is the perfect engine swap for mini-trucks, since it will generally slip into the same space as a four-banger. Street Rod fabricator Don Kendall used a stock '76 odd-fire Buick and Turbo 350 automatic to power his "shop truck," a fire-red '73 Datsun with lots of louvers and a lowered lid.

Craig Railsback, one of the owners of Blower Drive Service in Whittier, California, is a believer in V-6 power. His daily driver is a 1932 Ford pickup with a 4-71 GMC blown Buick under the hood. Besides the BDS blower kit, the engine features a Holley 4-barrel, Venolia pistons, homemade headers, and a remote oil filter.

If you really want to light a fire in a little pickup, install a V-6 and pump some horsepower into it. The "Country Butcher's" LUV stepside has a highly modified '81 Buick packing approximately 300 horses under a Mooneyham 4-71 blower, which is twisted by a Don Hampton drive kit.

The latest version of the Buick V-6 is the 3.0-liter engine, designed for transverse mounting in the '82 Century and other GM A-body cars. Although reduced in displacement, the short-stroke, high-revving motor produces the same horsepower as the 231 Buick.

Obviously the marketability of a 198 cubic inch V-6 went out the window. Buick increased the bent six's displacement to 225 cubic inches (3.75-inch bore by 3.40-inch stroke) in 1964, at the same time revising the bellhousing casting at the rear of the block to accept current-style Buick/-Olds/Cadillac transmissions. Even so, the V-6 was considered a dead player—or at least an anachronism—at the time, and Buick finally put the motor out to pasture. It was sold, patterns, tooling, and all, to Kaiser Jeep (later AMC) in 1967 for use in four-wheel-drive machines. Jeep used the motor only through 1971, when they switched to an inline six.

REINTRODUCTION OF THE V-6

Over the next ten years the political and energy climate in this country changed drastically. When gasoline prices and clean air regulations began to balloon in the 70's, General Motors first put their hopes in the all-aluminum Vega 4-cylinder. When it proved to be a limited success, they called up the old Chevy II cast-iron four-banger. Then, in 1975, Buick was able to buy back the V-6 from Jeep, and they jumped at the chance.

With no significant changes other than to bore the cylinders another .050-inch so it could share the same pistons with Buick's 350 V-8, the 231-inch "odd-fire" V-6 was immediately dropped into Buick Skyhawks and Oldsmobile Starfires, coupled to either a Turbo 350 automatic or Sag-

to performance when the hot rodders go to work with their big valves, hot cams, high-compression pistons, multiple carbs and the works? As for the V-6 engine in general, there is little doubt that we're going to see a lot more of it in the next few years. As cars get smaller (and there's no question but what the overall design trend is toward more compact dimensions) the size of the powerplant is going to become more critical. The V-6 layout appears to be one of the best compromises between size, cost, performance and fuel economy." Sounds surprisingly familiar, doesn't it? Except that such thinking—though accurate—was fifteen years ahead of its time at that point. With an upturn in the economy, and 29-cent gasoline still plentiful, American car buyers dumped the penny-pinching compacts and raced to buy behemoth-motored muscle cars. As demand for "economy" cars faltered, Detroit wisely stuffed the downsized bodies with oversized engines and called them factory hot rods instead.

As a Buick engineering and design exercise this 1980 Skylark X-car was treated to some fiberglass restyling and the transplant of a Riviera turbocharged 231 Buick V-6 by Doug Roe. With no engine modifications other than a water injector and a variable-boost regulator (which increased manifold pressure to 15 pounds), this car was reportedly capable of mid-14 to high-13-second quarter mile times.

inaw cast-iron 4-speed. The next year it was spread over the entire range of Chevy, Buick, Olds, and Pontiac small- and medium-sized cars, and by '77 an aluminum, 5-speed, overdrive, manual trans was also available.

When Buick decided that they were going to use the V-6 throughout their product line, including "luxury" cars like the Riviera, they knew they were going to have to do something about the vibration problem in the odd-fire motor. Minor though it was, the uneven firing pulse of the V-6 would generate complaints from Buick customers who were raised on nothing but silky-smooth eights since the 1930's.

There weren't many options. Buick was committed to the V-6, they were locked into the current design, and again they didn't have much time. The

Kenne-Bell, one of the nation's foremost Buick performance specialists, have recently been devoting most of their attention to the V-6. This wild 200-inch, fuel-injected, 10-second K/Gas Skyhawk is just one of their many "rolling test labs."

ingenious solution was to divide each of the three crankshaft journals in two, and stagger them 30 degrees with a narrow flange in between, to produce in actuality a six-throw crank in the 90-degree block without changing much else in the engine. The combination allows the pistons to arrive at top dead center in even intervals (120 crankshaft degrees), thus producing an "even-fire" 90-degree V-6. As noted in the introduction, the split-throw concept wasn't new, having been used by Lancia years ago and even by Caterpillar in large diesels, but it is certainly a new quantity to be reckoned with by today's performance engine builders.

The even-fire 231 was put into

production in mid-'77. By model year '78 Buick emphasized its performance commitment to the V-6 by installing a factory turbocharger system as an option (with 2- or 4-barrel carb) on Regal and LeSabre coupes. Later this system would find its way into front-wheel drive Rivieras as well. Also in 1978, a smaller 3.2-liter (196 cubic inch) version of the V-6 was added, primarily for use in Centuries as an economy option. The only significant difference in this engine was a decrease in bore size to 3.50 inches. It cannot be bored to 231 dimensions (3.80 inches), and currently has limited performance application, so be sure to avoid this engine if you are junkyard shopping for a Buick V-6.

For 1979 the performance potential of the V-6 was significantly improved with a new cylinder head design featuring taller intake ports and raised exhaust ports. To go with the new heads were a new intake manifold, carburetor, exhaust manifolds, and slightly hotter cam. Through these changes, stock horsepower for the '79 models was bumped up from 105 to 115.

In 1980 Buick once again upped the displacement of the V-6 with an optional 4.1-liter block (252 cubic inches) featuring a 3.965-inch bore. This engine, fitted in larger production models, comes with a Quadrajet 4-barrel carb on an aluminum intake and is rated at 152 horsepower.

For use in 1982 GM X-bodied cars (Buick Skylark, Chevy Citation, Olds Omega, Pontiac Phoenix) with front-wheel-drive and a transverse-mounted engine, Buick introduced a 3.0-liter version of the 90-degree V-6 which is

Just one of the several highly successful Buick V-6-powered machines competing in various forms of racing is this blown-fuel sand dragster. Built by Ted Ingersoll and now being campaigned by Bob Anderson as the Desert Pipeline Special, it has set several records on the 100-yard sand strip.

basically the same block as the 3.8-liter, but has a bellhousing shape to mate to the X-car transaxle. Unlike the '78-up 3.2-liter Buick, which uses the standard crank with a reduced 3.50-inch bore, the X-car block retains the standard 3.80-inch bore size and uses a destroked crankshaft (2.66-inch stroke length).

Whereas the 3.2-liter is simply an economy motor, the new highly "over-square" 3.0-liter engine should demonstrate some amazing rpm potential. This engine also comes with the high-port, late-model heads, and a "sideways" 2-barrel intake, a cross-over-pipe single exhaust, a starter swapped to the left (or "front") side, and a much shorter front engine cover. The 3.0-liter block will only fit an X-car chassis (a rear-drive bellhousing won't bolt directly to it), and although the crankshaft can be swapped into other Buicks, or vice versa, a major difference in the flywheel-attaching flange makes interchange of flywheels and clutches a major problem (for further discussion, see the clutch section).

BUICK V-6 PERFORMANCE PROJECTS

Obviously, the Buick V-6 remains in a constant state of upgrading, spurred on in no small part by a very aggressive

All 1964 and up Buick V-6's use the standard Buick, Olds and Cadillac bellhousing flange and starter mount, allowing a variety of transmission options for these motors. Also note balance weight and holes punched in the flexplate of this even-fire engine, accounting for partial crankshaft balance.

performance group in the Buick Division. A variety of performance test-car projects have been put together by Buick (including two Indy 500 pace cars—a Turbo Century in '76 and a 4.1-liter Regal in '81), or by independents such as Smokey Yunick, Jim Ruggles, Doug Roe, or Ray Baker working in conjunction with the factory. Several of these projects have been amply detailed in HOT ROD magazine, which has devoted considerable attention to the Buick V-6 in the last few years (several interesting Buick V-6 articles can be found in various 1978, 1979, 1980 and 1981 issues). Such projects have included a NASCAR-style, 231-inch, odd-fire, single 4-barrel motor producing 355 horsepower on gasoline and a supercharged 252-inch, 343-horsepower, street roadster motor, both built by Smokey Yunick; a 3400-pound '79 Century that ran 14.88/93.55 quarters with a reworked factory turbo and nitrous-oxide; a front-wheel-drive '80 Skylark X-car with a swapped-in 231 V-6 that ran mid-14's with nothing more than a factory-tweaked turbo; and an '80 Skylark with a 252-inch V-6 engine swap that dropped quarter mile times from 17.91/75.56 (with a stock, 173-inch, 60-degree V-6) to 14.97 seconds at 91.27mph.

In the last example, the Buick V-6 was modified with Smokey Yunick's "5-step" plan which includes a Weiand/Smokey 4-barrel intake and 600-cfm Holley, Weiand/Smokey headers, Crane H-214/2867-12 hydraulic cam and kit, and mild head porting. Such modifications are nothing more than the standard bolt-on tricks employed by most of today's street performance enthusiasts. In the case of the Buick V-6, according to Smokey's own dyno calculations, these modifications wake up the stock 152hp 4.1-liter V-6 to produce an impressive 230 horses.

There is no denying that the Buick responds excellently to traditional (and easy) bolt-on modifications such as cam, intake manifold, carburetor, ignition, and header changes, as well as a little mild head porting. When such a motor is slipped into a light car like an X-body or the earlier, slippery H-bodies (i.e., Skyhawk or Monza), mid-14-second quarter mile times in a very streetable, and still economical, machine are easily attainable. And a 14-second street driver is the type of car that means performance and fun, not downsized enthusiasm. What's more, 20-25mpg highway mileage comes with the package!

Since the Buick was the first V-6 on the American market, and since so much attention has been paid to its performance potential by both the factory and the press, it follows that the aftermarket parts manufacturers have been equally attentive to its needs. Such is certainly the case. Unlike other V-6 builders, owners of Buicks can choose from a wide variety of performance products ranging from cams to blower kits.

Although most parts houses offer Buick V-6 pieces, one company in particular specializes in Buick performance, with special emphasis these days on the V-6. Kenne-Bell Enterprises has devoted its entire attention to Buick performance since the company's founding in 1968.

Primarily involved in drag racing, K-B has always based product development on testing in their own competition cars. As early as 1968 they were racing a '64 Special with a '67 V-6 which was turning 93mph in 14.88 seconds with a 2-speed automatic. In the mid-70's they campaigned a highly successful NHRA R/Stock '75 Skyhawk with an odd-fire 231. Using a stock 2-barrel carb, headers, and a stock-spec "cheater" cam, it ran 14.40's at 94mph. Their current stocker, driven by Mike Doyle in the new NHRA SS/X class, has been running high-12's at 108mph in a Skyhawk with a 231 using 13.1:1 pistons, roller cam, high-port heads, and a K-B 4-barrel intake.

Another recent K-B project is an all-out Pro-Stock-style Skyhawk designed for K/Gas class with an injected, destroked 201-inch V-6 capable of 10.80/123 times. Kenne-Bell-sponsored V-6 machines also include Ron Cosner's K/Gas Pro-style Skyhawk, which turns consistent 10.20's near 130mph, and the Ingersoll/-Desert Pipeline 231-inch blown fuel sand dragster which runs in the 2's on the 100-yard course (and has beaten blown alcohol, aluminum, big-block V-8's!). Kenne-Bell should be known to

Although all Buick V-6 blocks appear essentially the same, several minor changes have occurred, especially in deck shape and water passage arrangement. At left is one of the few 3.8-liter heavy-duty blocks made in 1979, which has a deck similar to the "interim" block described in the text. Note round water holes above the cylinders and enlarged deck "ears". At right is the production 4.1-liter block which has an extended deck above the cylinders, four "D" holes, and other water passage changes. Also note differences in webs between lifter bores and the distance between bores on the two blocks.

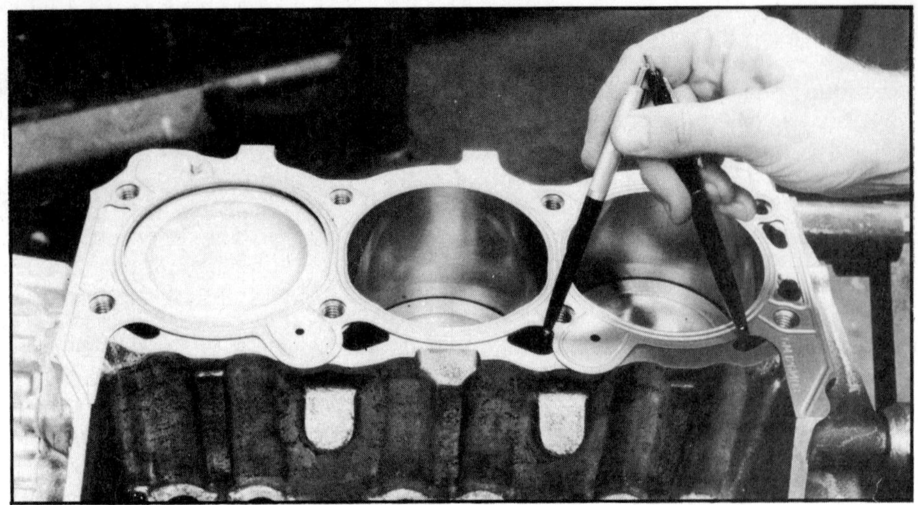

Although any Buick V-6 head will physically bolt onto any block, the changes in water passages and deck size can lead to substantial water leaks, as demonstrated by the obvious mismatch between this 4.1 block and an early head gasket.

anyone building a Buick. They not only have the most comprehensive line of performance products for the Buick V-6, but they offer a monthly newsletter and a tech-info telephone hotline as well. For information write to Kenne-Bell Enterprises, 1527 "K" W. 13th. St., Upland, California 91786, or call (714) 946-7671

The several examples of Buick V-6-powered machines listed above should not only indicate widespread interest in Buick V-6 performance, including welcome factory support and involvement, but also significant success in making the little V-6 fly. Full-bodied drag cars that can run in the low-10's on gasoline or mildly modified street machines that can turn honest mid-14's on the strip (even with front-wheel drive) command considerable respect even in the ranks of V-8-powered machines. When you realize that downsized cars fitted with V-6 engines ranging in size from 230 inches to 250 inches are very realistically capable of such performance, there's little reason to mourn the passing of big V-8's .

CYLINDER BLOCK

Pre-'64 Buick V-6 blocks have scant applicability because they used a completely different bellhousing and starter-mounting design than all subsequent versions. These early models have a more or less "round" bellhousing flange, and the starter bolts to the bellhousing rather than to the block. This 198-inch block has a bore of 3.625 inches, which could very likely be bored 1/8-inch over to match that of the '64 225 (3.75-inch bore), since these early blocks were not of current thinwall design. But given the scarcity of these engines, their condition after twenty years, and the limited bellhousing and transmission combinations available for them, they're not worth discussing except in terms of straight restoration. If you want a smaller displacement V-6, consider the 196-inch version introduced in 1978 or the new 3.0-liter X-car motor.

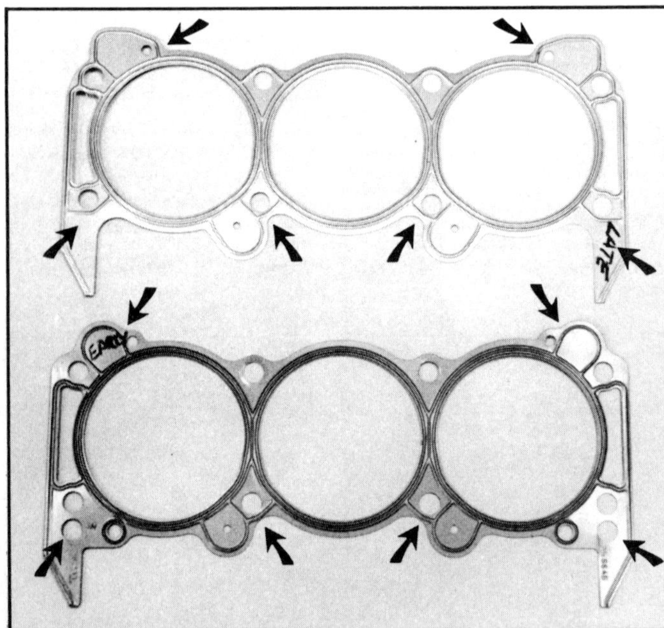

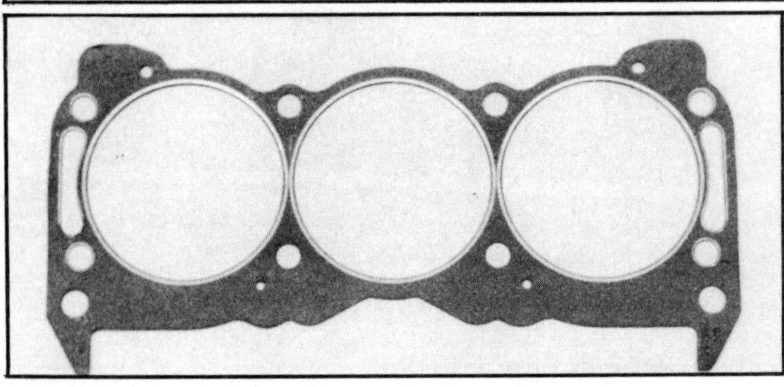

Besides other characteristics, different Buick blocks have identifying markings embossed on the bellhousing flange: 231 on left, 4.1-liter on right. The left block also has "HD" embossed on the back to denote that it is heavy duty.

The various head gaskets reveal Buick V-6 deck changes. Differences in the 4.1-liter (bottom) are apparent, but those between pre'77 (middle) and '77-79 (top) are more subtle—compare areas indicated by arrows.

When displacement was increased to 225 inches with both bore and stroke revisions in 1964, the block was also changed to the current Buick, Olds, and Cadillac bellhousing and starter design. The engine remained in this stage (3.75-inch bore) through its tenure at Kaiser-Jeep.

After Buick repurchased the V-6, the only significant block changes made before its reintroduction in '75 models as a 231-incher was a slight bore increase of .050-inch and an increase in diameter of the number 2, 3, and 4 cam bearings to make them all the same size (1.785-inch). Some revision of coring and water-jacket design was also done to reduce the weight of the engine.

WATER PASSAGE CHANGES

For model year 1977, cylinder head castings were changed to reduce weight and NOX emissions. At the same time some water passages were changed. The water-transfer holes from head to block located between combustion chambers on the intake port side were enlarged to a D-shape, and the cast boss surrounding them was correspondingly enlarged. Since the head surface on the Buick V-6 block closely follows the outline of the cylinders and water passages, material had to be added to the block to seal around these new water holes in the heads.

Using '77 or later heads (including the desirable high-port heads) on earlier blocks will cause significant water leaks between head and block. This is a very important point to remember when shopping for a Buick block to build for performance use, since you will probably want to use the late heads.

The 1975 and early 1976 231 blocks (casting number 1249432) have round water holes above the middle cylinders, with a sealing surface boss closely following the outline of the hole. They use head gasket number 1357869. In late '76 and early '77 two interim blocks were used. Casting number 1255860 was identical to the earlier block, except that extra block material was added around the round water-jacket holes above the cylinders to seal the larger D-shaped holes in the new heads. Casting number 1255862 was the same except that the dipstick entry location was moved from the left deck face to just above the left oil-pan flange. Mid-'77 through 1979 blocks (casting number 1254083) feature larger D-shaped water holes above the cylinders and a slightly extended deck to match the later heads. Either heads will fit the interim blocks; only '77-79 heads will fit the subsequent block without causing water leaks. The '77-79 heads require gasket number 1254483.

According to Jim Bell, another water passage revision was made in blocks and heads in 1980 on both the 3.8 and the 4.1 blocks. As a result, '79-type heads won't interchange on '80 blocks and vice versa. Unfortunately, this gets rather confusing, and even Buick experts seem to have lost track of exactly what fits what.

Remember, however, that any Buick V-6 head, from '61 to date, will physically bolt onto any V-6 block. The differences that preclude interchange are in the water passage size and location, plus the size and shape of the deck surface in some cases. To avoid problems, stick with matched heads and block whenever possible (i.e., if you want high-port heads, use a '77 or later engine). If you are swapping heads, you can check block-to-head compatibility by very carefully checking the head gasket match to the block and the heads. With embossed-steel shim gaskets, the exact size and location of gasket water holes is not as important as alignment of the embossed lips (raised edges) with block and head water passages. If the heads you want to use don't match your block, you have a couple of options which are discussed in the cylinder head section.

When the even-fire engine was introduced for model year 1978, no changes were made to the cylinder block. Consequently a '75-77 engine could be converted to even-fire by installing the split crank with compatible rods, pistons, cam, distributor, flywheel, and torsional dampener; but such a swap is probably not worth the effort unless you are shopping for a replacement block for a severely damaged even-fire motor.

For 1980 Buick offered a 4.1-liter (252-cubic-inch) version of the V-6, retaining the same crank but enlarging bore size to 3.965 inches. Again, 231 blocks cannot be bored out to accept the 252 pistons. Most (though not all) 4.1-liter blocks can be identified by three vertical ribs cast in the front between the timing cover and the manifold, plus "4.1" embossed on a rear flank near the bellhousing. On both 4.1- and 3.8-liter engines for 1980 the oil-suction holes in the block, as well as the oil pickup tube, have been enlarged to 5/8-inch.

BLOCK DESIGN SHORTCOMINGS AND STRENGTHS

When Smokey Yunick embarked on a high-performance engine build up project—one aimed primarily at possible NASCAR use—with the Buick in

Comparison of early heavy-duty block (left) with production 4.1-liter block shows significant strengthening in the pan rail and main-bearing bulkheads. However, note small diameter of oil pickup hole in the early block, compared to 5/8-inch oil passage in '80 and later production blocks.

Here's a better look at the thick bulkheads and pan rail in the heavy-duty block. This all-out race motor used in the Kenne-Bell gasser has also been fitted with K-B main-cap studs and fully-grooved main bearings.

This is a bottom view of the prototype 4.1-liter "heavy-duty" cylinder block. Note that the bottom end is similar to the earlier 231 heavy-duty block. However, it has larger oil pickup passages to improve oil evacuation from the sump.

late 1977, he had several unflattering initial observations about the cylinder block. He noted that the block (remember it was designed in 1961, and was taken directly from patterns intended for lightweight, aluminum castings) was unnecessarily heavy, having an extended skirt below the crankshaft centerline and plenty of cast-in external bosses and gussets. On the other hand, internal gusseting was lacking in the cam/main bearing bulkheads and between lifter bores, limiting block strength in these critical areas and leading to possible flexing and cracking. Furthermore, large water holes in the deck next to head-bolt holes weakened this area and increased chances of warping. Since the Buick V-6 design features only eight attaching bolts per head, maintaining good cylinder sealing with higher compression might be marginal. Smokey also felt the existing block and heads had too much water jacket area, yielding more than desirable heat (power) loss to the coolant.

Further, Smokey noted: (1) that the blocks were available only in 2-bolt main-cap versions; (2) that lifter bores were too short to maintain oil passage sealing with high-lift roller cams; (3) that the cam-bearing journals were very small, which limited maximum lobe height and therefore required a smaller base-circle diameter on high-lift cams (decreasing overall camshaft

torsional strength); (4) that the lifter bore spacing on the number 3 cylinder was wider than any others (causing possible roller alignment problems), and (5) that the open valley chamber not only reduced block strength but allowed oil drainback to fall on the spinning crank. In addition he decried the stock oiling system with its small passages, numerous right-angle bends, and circuitous route to the main bearings.

Such conditions begin to sound terminal, but such is certainly not the case. Admittedly the Buick V-6 was designed during an economic crunch to power compact passenger cars. It was meant to be an economy motor, not a racing powerplant. But that has never stopped hot rodders, who delight in turning grocery-getters into go-getters. Furthermore, the strong advantages of the 90-degree V-6 design—compact size, light weight, good manifolding, and excellent potential power to size and weight ratio couldn't be ignored.

Besides, for years it was the only game in town. Although the newer V-6's may be more refined in terms of performance potential, the Buicks are vastly more plentiful (therefore much less expensive in wrecking yards), and come in considerably larger displacements than the 60-degree engines. Furthermore, good, inexpensive bolt-on performance goodies abound for

the Buick. So don't let all this talk about oiling problems and outdated design discourage you. Currently, the Buick is by far the most practical V-6 choice for typical street or moderate racing performance. The deficiencies noted by Smokey Yunick, and proclaimed in many Buick V-6 articles, pertain only to the most extreme high-performance situations, such as Indianapolis, endurance road-racing or circle-track use, or all-out drags where the engine will see harsh conditions well over 6000rpm.

So if you are going to be modifying a production Buick V-6 with the usual additions of headers, ignition, cam, manifold, and carb, there's no reason to yank the thing out and tear it down to rework the block. Since the oil pump is on the outside, you can bolt on a higher volume or higher pressure unit (more on this later) to aid oiling. If you are going to have the engine apart anyway, it would be a good idea to install fully-grooved main bearings (available from Kenne-Bell, part KB12802), to radius the 90-degree bends in some of the oil passages, and to install a larger-diameter oil pickup tube on pre-1980 motors. Otherwise you can leave the block as is for the majority of street and "recreational racing" applications.

HEAVY-DUTY BLOCKS

For extreme or sustained racing,

however, you would be much wiser to start with one of Buick's heavy-duty blocks, none of which has been offered in production vehicles. The first (part 25500003) was produced in a limited run of approximately 200 units in 1979. This block came with partially finished cylinders featuring a 3.797-inch bore with unchamfered tops to allow finish honing to 3.80-inch size. Overbore possibilities were limited to approximately .030-inch (as on production blocks). Main bearing web thickness on this block was significantly increased, as was the thickness of the decks to reduce warpage. Whereas head bolt holes penetrate the water jacket on production blocks, they are blind on this heavy-duty block, and it can easily be identified by running a probe into one of them.

This block retains the production 2-bolt main caps, which should be aug-

mented by the addition of a main-bearing girdle for racing. Ted Ingersoll's nitro-burning blown sand dragster would regularly split blocks between the number one main and cam bearings, but never experienced bottom end failure using a cast production crank and a Kenne-Bell-girdle. Switching to the heavy-duty block, when it became available, solved the block-splitting problem.

Although a few 4.1-liter heavy-duty blocks were reported in the press as

being available in 1981, these were unique castings made for Buick experimental use only, primarily in conjunction with the Indy pace car project. However, two new heavy-duty blocks are currently scheduled for release in Spring of 1982. Specific details had not been completely finalized by our press deadline, but this new "Stage I" block is scheduled to come in two sizes, a 3.8-liter (part 25500011) with water between the bores and a 4.1-liter (part 25500015) with siamesed cylinders. There is a chance, however, that only one version of this block will be produced. It would feature siamesed cylinders and 3.80-inch bore size, with capability for overboring to the 4.1-liter dimensions (3.985 inches maximum—boring larger than this is prohibited on any 4.1-liter block because of the center-to-center cylinder spacing).

This prototype 4-bolt steel main cap was displayed by Diamond Racing for the new heavy-duty block. They should be available by the time you read this.

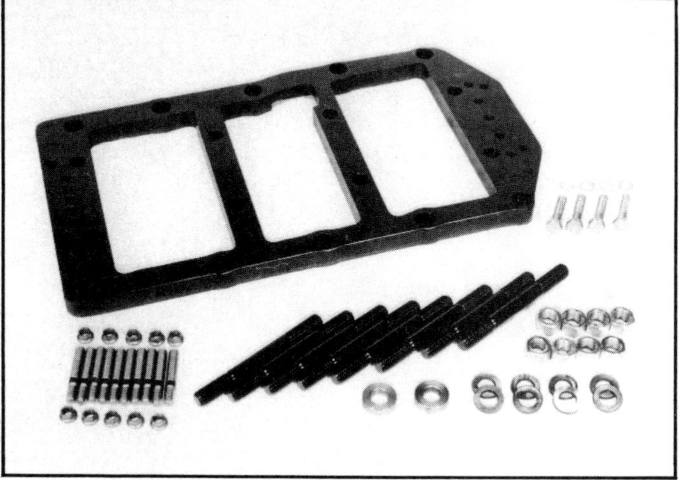

Steel block-reinforcing girdles are available for the Buick V-6 from Kenne-Bell (shown) and Moroso.

Besides thicker main-bearing webs and deck surfaces (though it will probably have open rather than blind head-bolt holes, to allow better coolant transfer), this block will also feature: a lifter gallery with reinforcing ribs; thicker (and possibly longer length) lifter bosses; increased pan rail structure; and reinforcing ribs on the rear face around the rear main bearing. Besides other visual changes, this block can be readily identified by changes in the water-passage holes in the deck face: the long water holes at the front and rear of the face have been eliminated, with two small round holes at the rear only instead. The prototype block we saw still had the large D-shaped holes above the cylinders, which may have to be filled and redrilled to match smaller dual holes in heavy-duty heads. These changes, plus the siamesing of the cylinders, should greatly increase deck strength and stability. This block will supposedly accept production heads, though it is designed to be used in conjunction with new heavy-duty heads, which are the most practical on a racing engine.

The oiling system in this heavy-duty block remains the same as that in production motors, except that oil passages to the pump are enlarged to 5/8-inch (as in late production motors). However, a "Stage II" version of the heavy-duty block is already on the drawing boards, which could become available as soon as mid-'82. This block would have a completely revised oiling system, with 1/2-inch diameter right and left main galleries spaced away from the lifter bores (so they do not intersect). As planned, this system would come partially machined, allowing the builder to tailor oiling to his specific needs. The main galleries would be drilled through the block, but plugged at each end so that an external pump (dry- or wet-sump) could be plumbed directly to them from either end. The builder would also have to drill individual holes (in whatever size and location he chooses) from the main galleries to each lifter bore.

The Stage II block may come machined to accept 4-bolt number two and number three main bearing caps, which are currently in prototype at both Diamond Racing and Kenne-Bell. These caps, which should be available by the time you read this, will fit any of the heavy-duty blocks after machining of the main webs to accept them.

The other alternative is to use a main-bearing support girdle, which is available from Moroso (part 3850) and from Kenne-Bell (part KB32201). The Moroso girdle fits over the stock main bearing caps, using shims to space it from the block rail. The Kenne-Bell girdle is a bit thicker, requires flat machining of the crown of the main caps, and sits flush on the pan rail.

The K-B girdle was designed in 1975, well before the introduction of the heavy-duty blocks, and it is intended to strengthen the entire block by tying together the lower end. It will reportedly reduce main web cracks, spun cam bearings, maincap walk, and even cylinder bore distortion on production blocks. Jim Bell highly recommends it for any block used for circle-track, heavy drag-racing, or any application requiring 6000-plus rpm engine speeds.

If you are using a production thin-wall block, the addition of a girdle (around $160) might be a more economical solution than starting over with a new heavy-duty block (about $450). However, installation of the girdle on a production block, which can be easily "bent out of shape," requires tedious and precise alignment (either by machining, shimming, or both) to make sure the girdle does not distort the block when installed. Obviously all align boring, cylinder honing, etc., should be done with the girdle (as well as honing plates and even the front cover) attached and torqued in place.

Other than oil system modifications, which we will discuss next, any V-6 block intended for serious racing should of course receive the usual detail preparation and blueprinting expected of any high-performance engine.

OIL SYSTEM

Although experts have disagreed on the best way to oil a Buick V-6, they all agree on one thing—the oil system in this engine is far from optimum *for critical high-rpm performance use.* Pay attention to the emphasis. You don't need a side-oiler on your Buick V-6 for street performance or weekend racing. Many builders say you don't need exotic oil system modifications even for more serious drag racing, such as in Super Stock or Econo-Dragster classes.

UNDERSTANDING THE BUICK OIL SYSTEM

What's the problem with the Buick

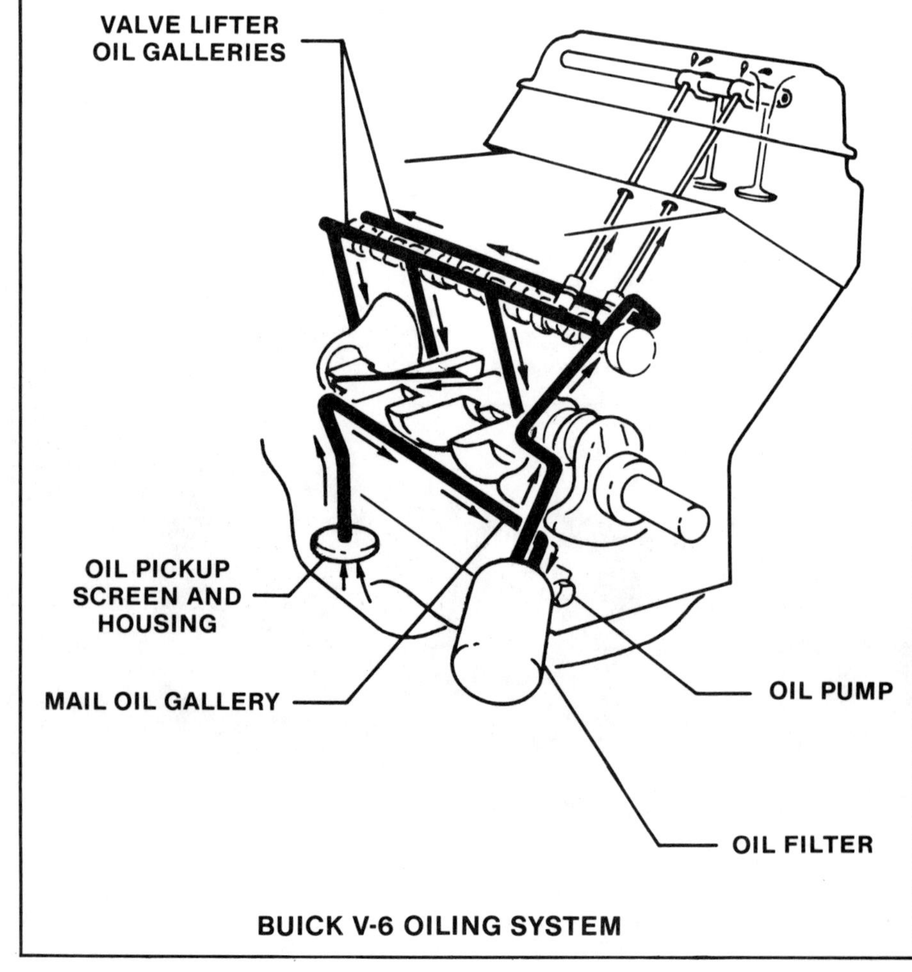

BUICK V-6 OILING SYSTEM

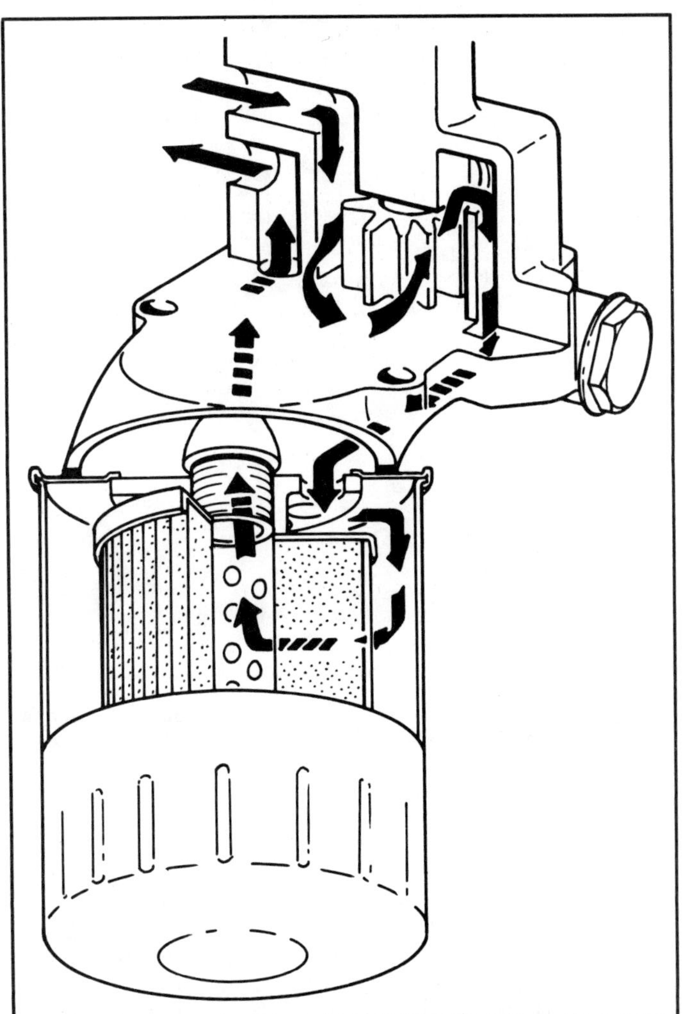

It takes some effort to understand how oil flows through the Buick system, especially if you don't have the components in hand to examine. Pictures help. On the back of the front cover, hole #1 matches the pickup passage in the block (leading to the pickup tube in the pan). Hole #2 does nothing and fits against a blank wall on the block. Hole #3 is the output passage from the pump, feeding into the pressure passage in the block. If you convert to an external oil feed into the block, plug this hole.

On the oil pump face there are three orifices. Note that #1 is the suction passage which feeds oil to the pump gears from the pan (through the block). The #2 passage is the output to transfer pressurized oil from the gears to the oil filter housing. The #3 passage matches the outlet from the filter housing and carries pressurized—filtered or bypassed—oil back into the block to lubricate the engine.

oil system? Let's take a tour through it to find out. We start in the middle of the oil pan, where suction from the pump (which is located on the right-front corner of the engine) draws oil through a pickup and relatively small (.50-inch in pre-'80 engines) pickup tube to a vertical 7/16-inch passage (pre-'80) in the middle of the right pan rail. This passage makes a 90-degree bend, runs horizontally to the front of the block, passes through a hole in the front cover, then down another 90-degree bend to reach the suction side of the oil-pump gear housing. The oil enters the gears from the top, passes between them, then exits through a slit on the other side into the oil filter housing. All of the oil should then pass through the filter, but since it is rather small some is usually diverted by the filter bypass valve (in the cast aluminum filter housing) directly into the outlet passage. The oil that does go through the filter re-enters the housing through the screw-on nipple to join the outlet passage. However, this outlet passage is also intersected by an opening to the oil pressure-relief valve, which is also contained in the filter adapter housing. When oil pres-

sure in the outlet passage (i.e., in the engine) builds to a certain level the pressure-relief valve opens, diverting oil from the outlet passage directly back into the suction side of the pump gears (from the bottom).

The oil that is pumped out of the exit orifice of the filter adapter housing re-enters the pump housing, makes another 90-degree turn, passes through the engine front cover into the block, then turns right once again to follow a

passage up to right-hand lifter gallery. In pre-'80 blocks, all of these passages are 7/16-inch in diameter. In the Buick V-6 the right-hand lifter gallery (also 7/16-inch diameter) is the main engine oil gallery. It intersects all six right-side lifter bores, plus feeds oil to the cam bearings, and goes on to the main bearings, through passages approximately 1/4-inch in diameter. A groove in the front cam journal feeds oil through another small passage to the

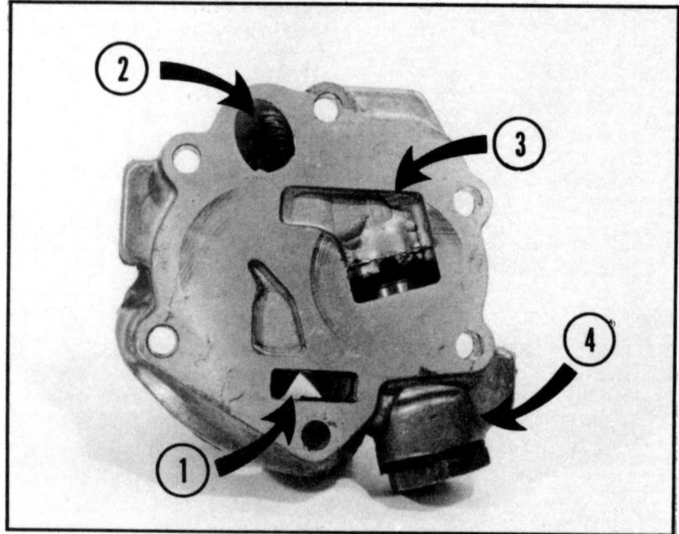

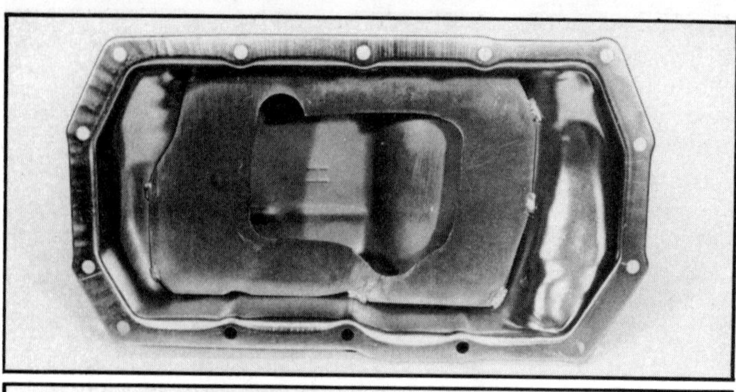

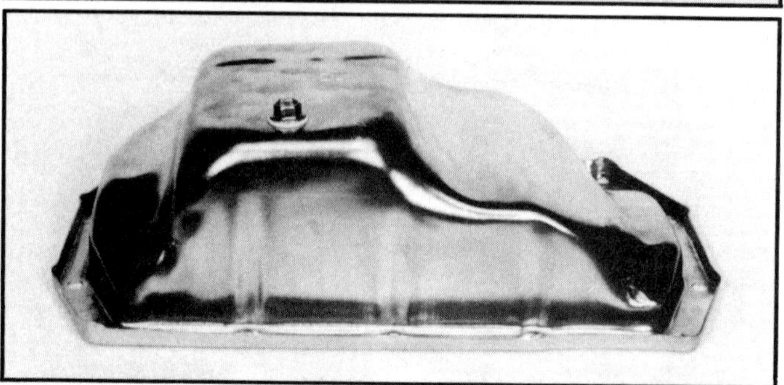

The filter housing, which bolts against the bottom of the pump, also includes the oil pressure bypass valve(4). Passage #1 matches the outlet passage from the pump and feeds oil to the filter. Passage #2 is the outlet from the filter, into which the filter bypass feeds. It matches with passage #3 in the front cover. The pressure bypass valve, which bleeds off excess oil from the filter outlet and/or filter bypass, feeds it back into the pump gears through opening #3.

The Kenne-Bell #19904 oil pan with a welded 360-degree baffle is popular for wet-sump systems, and fits the Skyhawk or Monza chassis.

left lifter gallery, which is also 7/16-inch in diameter. Pre-'75 engines also have small passages in the front of the block running from the lifter galleries to the front rockershaft pedestals in the heads to lubricate the cast rockerarms through the shafts. All '75 and later engines, using stamped steel rockerarms, oil through the pushrods (more on this in the valvetrain section).

The deficiencies in the Buick V-6 oiling system for high-performance use should be apparent—too many relatively small passages, far too many sharp right-angle bends, and too many places to lose oil pressure or volume before lubrication gets to the main and rod bearings, especially those at the rear of the engine.

OIL SYSTEM MODIFICATIONS FOR STREET PERFORMANCE

But what about street engines? Remember, Buick is running factory turbos on these things with the stock oil system, and they certainly aren't planning major warranty repairs. Again, opinions differ slightly over oiling a street or moderate-performance Buick V-6, but they do agree that complicated major oil system reworking is only necessary for radical or endurance racing engines. For street and moderate drag racing, Jim Bell strongly recommends one system—a Kenne-Bell high-volume oil pump kit (KB18202), which consists of 1/4-inch longer pump gears, a 1/4-inch pump housing spacer plate, a higher pressure-relief spring and shim, and all necessary bolts, guide pins, and gaskets.

One very nice thing about the Buick oil system design is that the pump is on the outside of the engine, so you don't even have to pull the pan to install the kit. Jim notes that stock late-model V-6's ('75-up) run very low oil pressure, about 35psi maximum and 8psi at idle. He recommends this oil pump kit even for street stockers. The kit increases oil volume about 40%. The stronger relief spring brings oil pressure up to 60psi; adding the shim in the kit increases it to 80psi.

The Kenne-Bell SS/X Skyhawk is a good example of the effectiveness of

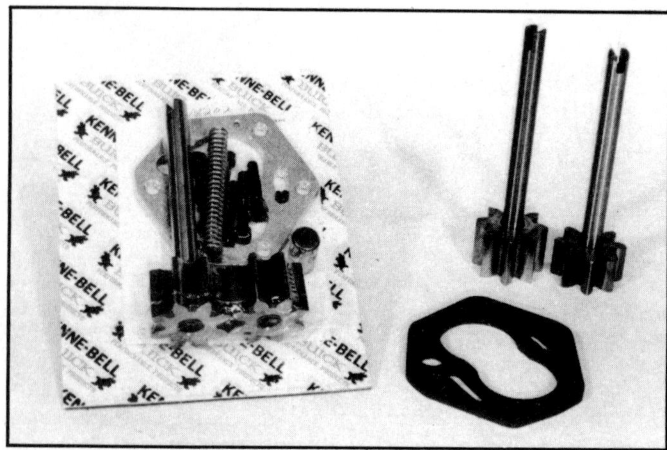

Kenne-Bell swears that a high-pressure, high-volume oil pump kit is all that most Buicks need. The kit includes longer gears (long gear is compared with stock gear at right), spacer plate, higher pressure-relief spring and necessary bolts.

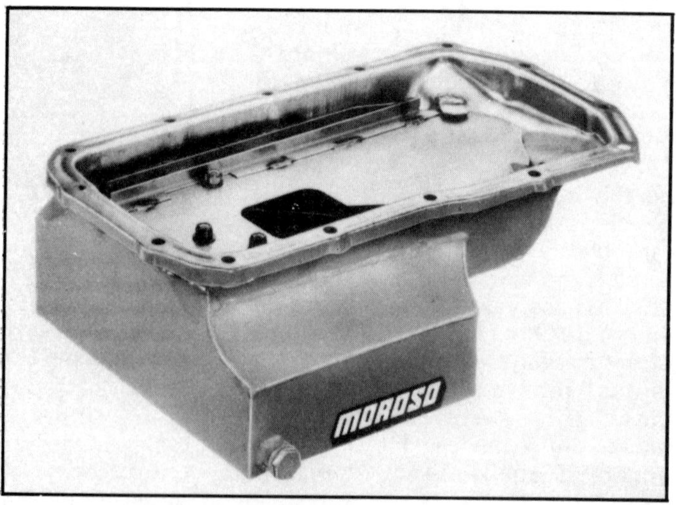

The Moroso road-race, wet-sump pan features full trap-door baffling plus a windage tray.

the K-B oil pump kit. The car has over 200 runs on it as of this writing without a failure. It leaves the line at 8000rpm and shifts at 7500. The engine has absolutely no modifications whatsoever to any of the oil passages in the block or oil pump housing. The only changes are the Kenne-Bell oil pump kit, a KB19904 oil pan, and enlarged pickup tube. This combination shows 80psi oil pressure in the traps (about 7500rpm).

Any time you dissassemble the oil pump on a Buick, it is wise to pack the gears with Vaseline before reinstalling the cover. Since the external pump is located so far from the oil pan, it often has a difficult time self-priming if the engine is started with a dry pump. Packing the pump is a good way to prime it, and thereby save new bearings and other engine parts from a dry start. Better yet, pull the distributor and use something like a large screwdriver blade chucked in an electric drill motor to spin the oil pump and circulate oil throughout the system before turning the engine over. Kenne-Bell also makes a special shaft for this purpose (KB40201).

OIL SYSTEM VARIATIONS

Besides some oil pan variations and a dipstick relocation (in 1977), there have been few changes to the Buick V-6 oil system. In 1975, a metric-threaded, oil-filter nipple was inserted in the filter adapter housing (oil-pump cover) to accept a relatively small-diameter AC PF40 metric filter. This pump housing is readily identified by the word "METRIC" embossed on the outside. Earlier housings are identical except that they take SAE-threaded AC PF24 or PF46 filters. The housings will interchange. Incidentally, whereas all other Buick V-6's have an angled mount for the oil filter (designed to clear the front crossmember in the original '61 Special), the aluminum 215 V-8, V-6's used in Cadillacs, and the new 3.0-liter version of the V-6 for use in X-car chassis mount the filter parallel (therefore closer) to the front of the engine. If you are swapping the V-6 into tight quarters (such as into an X-car), using a filter adapter from one of these engines will help.

Another pump housing variation can cause plenty of trouble for those who change oil-pressure relief springs. In 1979 Buick changed the filter housing to take a relief spring approximately .50-inch shorter than previous models. If an earlier spring is installed in this housing it will "stack up," affording no oil pressure relief at

all, possibly causing considerable engine damage. As of 1982, this short-spring housing is the only one available from Buick. If you are going to install an aftermarket high-pressure relief spring in your motor, make sure it is close in length to the one you take out. (Free length of the pre-'79 spring is 3.100 inches; of the '79-up spring is 2.550 inches.) The Kenne-Bell oil system kits now come with two replacement springs, one of each size.

Several sizes and shapes of oil pans and matching pickup tubes have been used on Buick V-6's over the years. Use as large a pan as you can in your chassis, such as part 25505552 (which takes pickup 25505644 with 5/8-inch diameter tube) to insure constant oil supply, especially with a higher volume pump. Besides having several internal oil passages enlarged to .50- or 9/16-inch, 1980 and later engines feature 5/8-inch diameter oil pickup tubes, which are preferable; however, stock pans to match will not fit in the popular Skyhawk (or H-body) chassis. An excellent pan that will fit this application (and most others) is Kenne-Bell's KB19904, which has a

A windage tray is cheap bolt-on horsepower for any engine. Kenne-Bell makes this one for Buick V-6 engines.

An easy way to reduce restriction in the Buick oil system is to adapt a remote, free-flowing, single or dual oil-filter assembly in place of the stock unit.

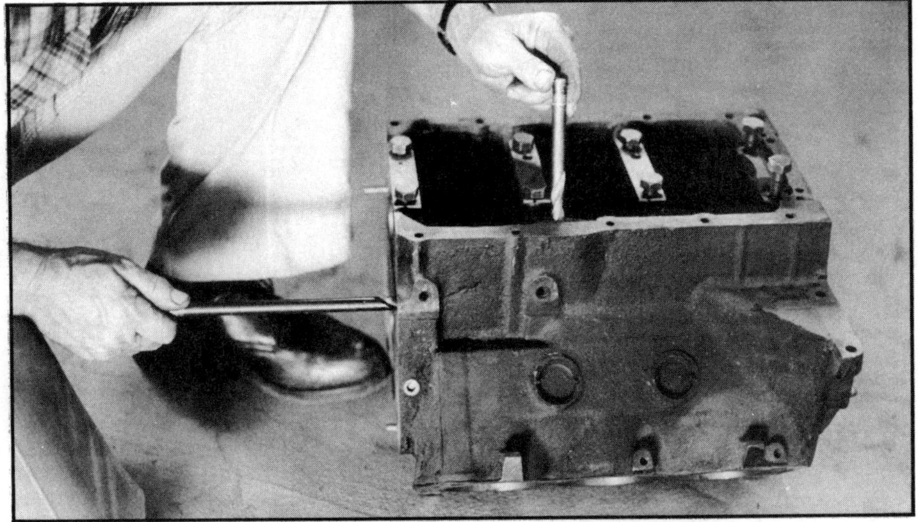

Doug Roe demonstrates how he uses long-shank 1/2- or 9/16-inch drill bits to open up the oil suction passages in pre-'80 blocks. The right-angle junction of the two holes should then be radiused with a ball stone on a long shank.

Part of **Jim Ruggles' modification to the Buick oil pump is illustrated on this stock pump housing and filter adapter. To keep bypassed oil from being dumped back directly into the pump gears, he adds the block-off plate shown, then he mills an opening between the bypass cavity and the incoming oil passage (arrow). Hard anodizing the aluminum block-off plate also helps it resist wear from the pump gears, a problem often encountered on the stock aluminum filter housing.**

welded-in, 360-degree baffle tray. It should be used in conjunction with their KB19801 windage tray kit. For this combination, it would be wise to construct your own 5/8-inch i.d. pickup tube. Kenne-Bell also lists a 1.5-inch-extended deep-sump pan (KB19906) which comes with extended pickup. Moroso also offers a 4-quart capacity, trapdoor-baffled, wet-sump, road-racing pan for the V-6 (part 2170) designed by Smokey Yunick. It comes with a bolt-in scraper/windage tray and takes extended pickup part 2466. Other than possibly adding a remote filter adapter to the Buick, the vast majority of street performance, recreational off-road, or weekend racer

Buick V-6's need no further modification to the oil system.

OIL SYSTEM MODIFICATIONS FOR HIGH PERFORMANCE

However, if you are going to be doing heavy or prolonged racing with a Buick, you are going to have to make some changes. Considerable attention has been given to this subject—

much of it conflicting. There are two basic approaches: modifying the stock-type pump and internal oil passages, or converting to a dry-sump with external plumbing. Plus, there are variations or combinations of both.

First, let's look at what can be done with the stock system. Beyond the upgrading already discussed, the next step would be to add a large-capacity, low-restriction, remote oil filter such as the excellent new Motorcraft FL-1HP. Offenhauser makes a simple screw-on adapter (available from Kenne-Bell and Moroso), which replaces the stock oil filter, to attach oil hoses for this purpose. Such a set-up utilizes the stock Buick pressure-relief valve and passages in the pump, but greatly reduces restriction caused by the small, stock filter, and increases the amount of oil filtered (rather than bypassed).

The next step would be to enlarge all major oil-feed passages in the block, front cover, and pump housing using long-shank .50- or 9/16-inch drill bits, and radiusing the several right-angle corners with a ball-shaped grinding stone. The operation is mostly logical—enlarge and streamline all oil suction and pressure passages up to (but not including) the right-hand lifter gallery.

Oil needs to be restricted in the passage to the left-hand lifter gallery, which is fed from a hole in the number

If you want to avoid a few right angles between the oil pump and the main gallery, you can tap into the pressure passage of the pump housing as shown in the left photo. If so, plug the hole indicated by the arrow. Then you can plumb this auxiliary oil line directly into the block through the hole in the right front corner (previously used for the oil-pressure sensor), as shown in the right photo. This setup was devised by Smokey Yunick for a wet-sump oil system. Also note the machined aluminum adapter in place of the stock oil filter to route lines to a remote filter.

one cam bearing. This feed can be effectively reduced by installing the bearing so that the oil holes partially cover the orifices in the block. Otherwise, you can drill new holes in the bearing (.135-inch on the feed side, .085-inch on the discharge side), and insert it so that these holes align with the block passages. Oil to the numbers 2, 3, and 4 cam bearings should be similarly restricted. Oiling to the right-hand lifter gallery obviously cannot be reduced, since this is also the main oil gallery. However, on '75 and later engines, which oil the top end through the pushrods, aftermarket restrictor-type pushrods (with smaller holes in the tips) will regulate top-end oil supply.

Jim Ruggles, who has done considerable performance V-6 building and testing for Buick, sees the stock pressure-relief system as another major culprit of oiling problems. "In the stock engine," says Ruggles, "100% of the oil coming out of the pump is directed through the filter—which we know it cannot do because of the small filter size. So in actuality, over the life of the engine, only 50-60% of the oil gets filtered." The remainder is diverted by the filter bypass valve into the outlet passage. The oil pump pressure-relief valve is also in the same outlet passage, so as pressure builds up in the engine, all excess oil (including that bypassed by the filter) is directed by the relief valve right back into the pump gears. As Ruggles explains, "The oil is bypassed on the way back into the engine, directly back into the pump. This builds up head pressure in the pump, especially on cold starts. You tear up the timing chain, the distributor drive gear, and so on."

Ruggles' solution is twofold. First, he modifies the stock pump with a special 1/8-inch spacer plate (see photos) to redirect the path of oil flow from the pressure-relief valve into the intake passage of the pump housing, rather than directly into the gears. At the same time he elongates the relief valve slot slightly to allow more oil to be bypassed. These changes reduce the amount of oil that actually passes through the gears, thus reducing strain on them and connected engine parts.

Secondly, Ruggles eliminates the restrictive stock oil filter entirely by replacing it with a hand-made aluminum cover plate which is drilled for one large outlet oil line only. He also plugs the stock bypass valve in the filter housing. Then he runs an outlet oil line from this adapter plate to a large remote filter, through a cooler, then directly back into the engine at the (enlarged) hole normally used for the oil pressure sender. This way oil pressure is regulated before it passes through the filter, less oil is actually run through the pump gears, and all oil coming out of the pump is filtered and cooled before entering the engine. This system has been used by Ruggles and Doug Roe on numerous IMSA road-racing Buick V-6's with success.

Once this modified oil supply is fed back into the engine, some further plumbing alterations are in order. Ruggles enlarges the oil passage from the entry point (oil sender hole) up to the right lifter gallery to 19/32-inch. "It doesn't go directly into the main oil gallery. It hits about 3/8-inch below, because on the production oiling system they went straight across and drilled into the opposite bank," explains Ruggles, "so you have to get in there with a little ball-faced grinder and radius all those corners, too." This area is accessible by removing the small soft plug covering the end of the lifter gallery in the front face of the block.

Most V-6 builders like to replace the soft plugs at the front of both lifter galleries with screw-in pipe plugs, since all oil pressure would be lost if a plug worked loose. However, a very thin plug is required so that it doesn't screw into the passage, restricting oil flow, or protrude from the face of the block, to interfere with the cam gear. Kenne-Bell now makes thin Allen plugs for this purpose (KB18105). When routing the oil-feed line externally into the block, as described above, the oil-feed passage from the front cover must be plugged in the block or in the front cover.

Ruggles also enlarges the oil passages running from the number two and number three main bearings up to

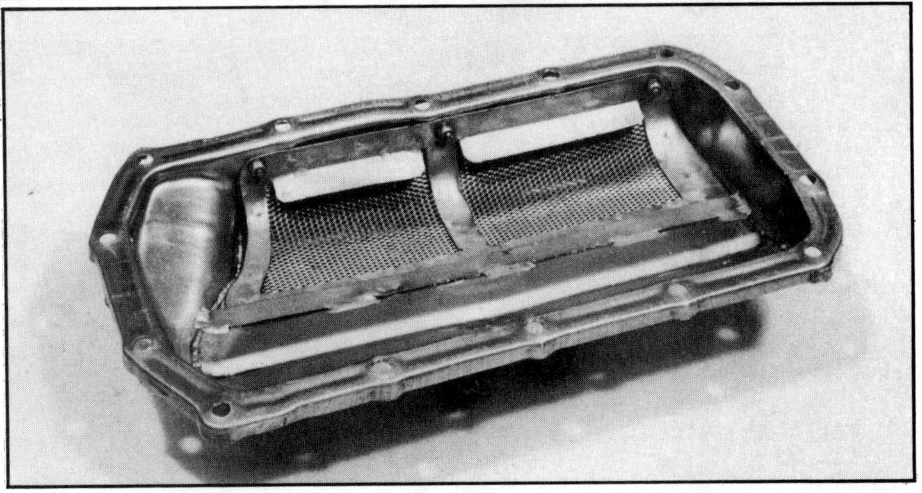

Using a dry-sump pan and belt-driven pumps will solve all the oiling problems in the Buick, up to the point where the pressurized oil enters the block. However, in most cases the actual benefits of such a system probably won't warrant the expense.

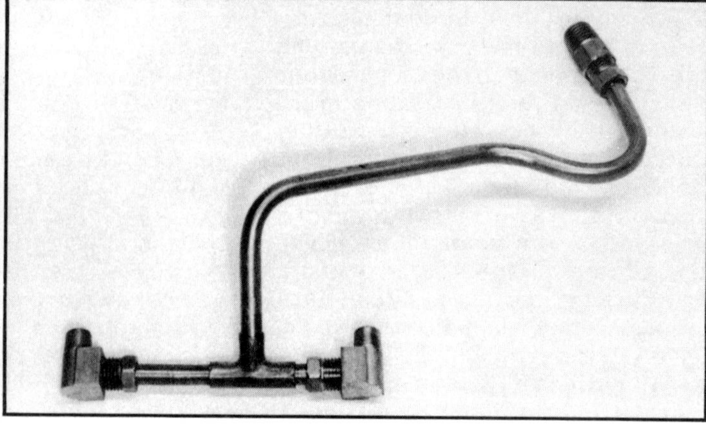

The Smokey Yunick side-oiler for the Buick really isn't very complicated. Basically it's a short tube to carry oil from the block oil-entry passage over to two new passages drilled through the side of the block, where it is supplied to the middle pair of main bearings. The only tricky part is to drill the new supply passages to the bearings in the proper location.

the right-hand lifter gallery to 3/8-inch, and he uses a clever trick to restrict oil to the cam bearings: "I drill a new hole (.135-inch) in the bearing shell and turn the original hole so that it's against a blank wall. The oil passage going down to the main journal intersects the cam bearing bore, and I try to position the oil hole in the cam bearing as close as possible to the bottom of that intersection and make the camshaft meter the oil out. It's good for 15 pounds of pressure."

As for enlarging block passages up to the oil pump, Ruggles had these comments: "I find that if you restrict oil losses in the engine enough, you don't need to go too big on the pickup passages. About .50-inch is enough. Some early blocks won't take 9/16- or 5/8-inch. I have broken through the sides of the passages on a couple. Early '80 and up blocks come with stock 5/8-inch passages."

"In order to prevent oil draining out of the pump, causing it to lose its prime when you shut the motor off, there's a 180-degree turn into the pump. The passage comes into the front cover from the block, makes a 90-degree turn, goes up to a 180-degree corner, then drops down to the pump. It's imperative that you make generous radiuses on all these corners, especially the 180-degree at the top. You can get at it from the back side, because the back is a blank wall against the front of the block. I get in there with a ball grinder and put a nice radius on it—make that as large as possible. Some guys will just go in with a 3/4-inch bore fluke mill and cut

The Kenne-Bell top-oiler feeds the mains from above, by tapping into the passages that run past the cam bearings. The right photo shows the small line that carries oil from the junction block to the rockershaft, so that the rockerarms are oiled through the shafts.

straight in to get a lot of area. But I try to get a nice radius."

Another possibility for oiling the Buick suggested by Jim Ruggles but not actually tried on a motor as yet, would be to mount a pressure-relief valve at the rear of the right-hand oil-feed (and lifter) gallery. This could be done fairly simply by removing the gallery plug at the back of the block, attaching a home-made relief valve (you could even use the stock piston and spring in a machined block of aluminum), and then running an oil line from the relief port in the valve directly down to the pan. Such a system would be very similar in concept to that used on side-oiler 427 Fords, and would insure positive oiling to the main bearings at all times, as well as full filtering of all oil. The stock relief valve in the Buick pump could simply be blocked shut, or else a new outlet for an oil line could be tapped into the pressure side of the pump. Then an oil

line could be run from the pump, through a remote filter and cooler, and into the engine as described above.

And, just as you can use an external line to connect the pressure side of the pump to the block, you could also fabricate an external oil line from a pickup mounted in the pan to a fitting tapped into the suction side of the pump housing. This would be a good way to avoid the right-angled block passages on the intake side of the system. Just remember to include something (such as a check valve) to keep the pump primed.

DRY SUMP AND THE "SIDE OILER"

The alternative to modifying the stock wet-sump oil system in the Buick is to convert the engine to a dry-sump system. This is the most practical approach for a serious racing engine, since it eliminates most of the trouble-spots in the stock system. (In fact, the only reason not to use a dry-sump for racing is the high cost of the components.) Aviad is the most common supplier of pans and other parts for dry-sumps, and individual pieces for the Buick V-6 are stocked by Moroso, among others.

With a dry-sump, oil is drawn directly out of the bottom of the pan through one or two large scavenge hoses, thereby solving the problem of the small pickup tube and block passages, as well as the sharp turns, leading to the pump in the stock system. And since a dry-sump relies on its own belt-driven external pumps and bypass system, the problems of the Buick pump, relief valve, filter, and internal passages are circumvented. The only remaining problem in the Buick is routing the oil from the pressure pump throughout the engine.

Smokey Yunick has used a much-discussed "side-oiler" system of external plumbing on the Buick. In the latest version he drills two small passages through the lower-right side of the block (below the water jacket, naturally) directly into the main-bearing bulkheads to the number two and three main bearing journals. These holes are tapped for pipe fittings on the outside of the block, and connected by copper tubing to a T-fitting where the main pressure oil line enters the block. As in other systems mentioned previously, Smokey uses the oil sender hole at the right-front corner of the block, enlarged to .50-inch, as the main oil entry location. This system consequently feeds oil through the block passages to the front and rear mains (as well as other oiling locations), while

The difference between odd- and even-fire Buick crankshafts is immediately apparent. Since the main-journal size and spacing is the same on both, they can be interchanged—though there is little advantage in doing so.

diverting a separate subsystem directly to the middle two main bearings.

Smokey uses the side-oiler for both wet- and dry-sump oil systems on Buicks. Other than the precise drilling of the two passages through the main webs, it is a relatively simple system. Baker Engineering (Grand Rapids, Michigan) will install the Smokey side-oiler on Buick blocks at their shop or, if you have adequate machining facilities and experience, they will ship you a kit and explain how to install it.

Kenne-Bell, on the other hand, devised a "top-oiler" system for their V-6 gasser, similar to the type commonly used on Ford 351-Cleveland NASCAR engines. This system circumvents the stock oiling in the block altogether. It utilizes a manifold (resembling a fuel block for multiple carbs) in the valley chamber that feeds individual oil lines connected to holes drilled into each main-bearing web from the top. Two smaller lines run from this manifold to the front of each rockershaft to feed oil to the rockerarms through the shafts. Lifters are oiled by drip-down from the rockerarms and mist in the valley. This system also eliminates the problem of sealing the right-hand lifter gallery with special rollers at the same time that it controls top end oiling on both sides.

Whether one system is better than the other is a moot point. As Jim Bell says, "We came up with this design before the side-oiler thing, and it works fine, so we haven't tried anything else." Kenne-Bell sells the top-oiler kit, but notes that it requires considerable machining of the block (which they will perform).

Undoubtedly other good ways to improve the Buick V-6 oil system exist. The point is that there are several approaches to lubricating this engine for strenuous racing and that devising such oil system modifications takes nothing more than common sense and ingenuity (along with a good understanding of lubrication principles). And remember, similar oil system reworking is typical of most production engines modified for extreme levels of performance.

It is worth mentioning one more time: *such extensive oil system modification is only necessary for severe racing conditions.* Many racers—notably the Desert Pipeline blown fuel sand dragster—still rely on the stock oiling system left essentially as-is with no problems. Furthermore, major oil system modifications, especially for long-distance, high-rpm racing, are certainly not unique to the Buick V-6. So do not let all of this discussion lead you to think that the stock V-6 oil system is "junk," especially for the purpose for which it was intended—lubricating a street-driven engine under normal operating conditions.

CRANKSHAFT

Other than the first 198-inch version of the Buick V-6, which had a stroke of 3.20 inches and main-bearing-journal diameter of 2.30 inches, and the new 3.0-liter Buick, which has a stroke length of 2.66 inches, all engines share the same stroke of 3.40 inches and main-journal diameter of 2.50 inches, so crankshafts are interchangeable in all Buick V-6's from '64

All Buick V-6 cranks have rolled fillets on main journals, but only the even-fire turbo crank (used in factory turbo-charged engines) has fully rolled fillets on all rod journals as well (bottom), adding considerably to the strength.

up.

The big change came late in 1977 when Buick introduced the 30-degree split throw crankshaft to produce the even-fire V-6. We have described this design in some detail elsewhere, so we won't repeat it here. However, two versions of the even-fire crank have been offered so far. Part 1260873, which came in all turbocharged production motors, has rolled fillets on all main and rod journals (see photo). This seemingly small difference greatly increases the strength of this crankshaft; it is definitely the best choice for a performance build-up. Smokey Yunick, among others, now uses the stock, cast, rolled-fillet crank for performance engines, rather than far more costly and difficult-to-obtain steel billets. In general, race-prepped versions of the Buick V-6 have experienced very little crankshaft breakage.

In late 1981, Buick's Product Engineering department announced the intent to make available, among other things, an all-new, heavy-duty, forged-steel crankshaft, assigning it part number 25500008 and an availability date of May 1982. Preliminary description of this special-order forged crank says it will have maximum counterweighting, a 2.66- to 3.40-inch stroke capability, and will be sold "unmachined," i.e., neither main nor rod bearing surfaces will be finish machined and ground. (However, in the past some "special" cranks promised by manufacturers have never materialized. So, as with any such proposed special parts, we would suggest a "wait and see" attitude.)

The case for a forged crank for the Buick V-6 is not all that strong, anyway, since the cast crankshaft appears to be quite tough. The forging would certainly be a necessity if Buick were to make a determined effort at a stock-block Indy motor. Otherwise, the major advantage of this crank (besides additional strength) would be the capability to alter rod-journal width, rod-journal diameter, or stroke length.

EVEN-FIRE VS. ODD-FIRE

Even- and odd-fire cranks will obviously interchange in Buick V-6's, so you can easily convert one type of engine to the other. There seems to be little reason to do so, however, since advantages and disadvantages of either design are nearly equal. Power wise, there is virtually no difference between even- or odd-fire. This stands to reason since the firing order and manifolding remain the same in both versions. Viewed another way, if you think of the V-6 as two 3-cylinder inlines connected at the crankshaft (which it is), the individual cylinder banks operate exactly the same in either the even- or odd-fire configuration.

The odd-fire motor obviously has the drawback of uneven firing pulses, which may bother some people. More physically detrimental in a competition motor is the uneven firing sequence of the ignition system, which does not allow equal voltage build-up times in the coil between spark firings. A final consideration to performance is that all standard-shift, odd-fire motors come with a flywheel approximately twenty pounds heavier than that used in even-fire motors, to help absorb torque fluctuations (which means they absorb a portion of the engine's power output during acceleration).

Since rod-journal diameter was increased from 2.0 inches on the odd-fire crank to 2.25 inches on the even-fire, thereby increasing the "over-lap" cross-section of the split pins considerably, there seems to be no real strength advantage in the odd-fire crank. As a matter of fact, when testing the new split-throw crank for durability, Buick designers noted that crank failures were caused by bending loads placed on the crankshaft nose by accessory drives, rather than torsional stress within the crank. It was discovered at the factory, and born out by some racers, that crank failures in the Buick V-6 almost always occur between the number one and number two main journals. This can be at least partially attributed to the placement of all accessory drives at the front of the motor, plus the extra length thereby necessary in the crank snout. The drives place a twisting (torsional) load on the crank, while the extra length in the snout increases the bending load (moment) at the belt drive pulleys. To remedy the problem, the designers added some material in the crossover sections of the even-fire crank, especially between the number one rod journal and number two main journal. So the straight-pin, odd-fire crank really has no strength advantage over the even-fire, and might even be weaker (certainly more so than the rolled-fillet turbo crank). This point of concentrated bending loads on the crank snout should be given careful attention by all race engine builders, especially in applications that require additional accessory drives on the front of the motor (e.g., dry-sump pump or blower-drive pulleys.)

So the only real advantage of the odd-fire crank is that the rods and rod bearings can be positioned more near-

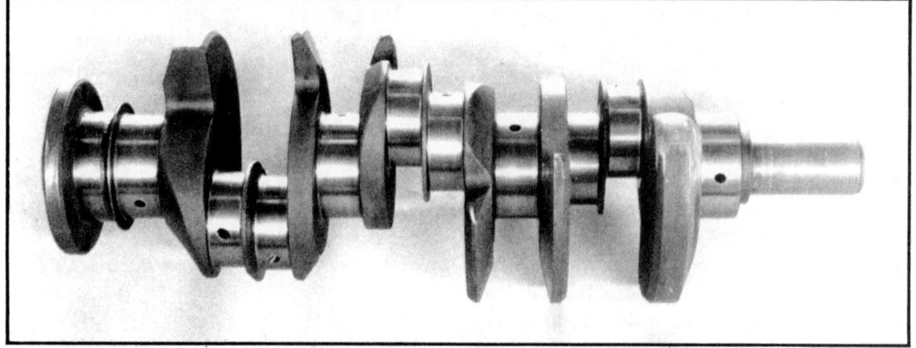

Since stock Buick V-6 cranks are quite strong and custom cranks are quite expensive, the only pratical reason for having a billet crank made (as shown) would be to increase stroke length or alter rod-bearing width and offset.

ly in the center of the piston bores, as intended from the original 90-degree V-8 design. Many people do not realize that in the odd-fire motor the centerline of (half) the connecting rod journal, and therefore the connecting rod beam, is already offset slightly from the centerline of the cylinder bore. This configuration is common to most 90-degree V-8 designs so that the stagger between cylinder banks can be reduced and therefore the overall length of the engine can be shortened. In the odd-fire Buick the amount of rod journal offset is .065-inch (see illustration).

On the even-fire crankshaft, however, each rod journal had to be moved over another .060-inch to allow for the .120-inch wide flange between adjacent journals. Given the fact that cylinder bore spacing was predetermined by the existing V-8 design, engineers had several options to accommodate the flange needed between the staggered crank-throws. They could: (a) narrow the rod journal width, (b) offset the rod bearing and the big end of the rod, (c) offset the rod beam in the cylinder bore, (d) offset the little end of the rod in the piston, or any combination of the above.

The Buick designers chose to retain the full width of the rod journal (unlike the 90-degree Chevy V-6, which has narrower journals than the smallblock V-8), and to retain basically the same connecting rod dimensions as the odd-fire. Consequently, in the even-fire Buick the rod beam is nearly centered on the rod journal, but is considerably offset (almost .100-inch) in the cylinder bore and in the piston.

Interestingly enough, potential problems caused by the offset rod in the split-throw, even-fire engine have seldom been discussed in relation to the Buick; the subject has come to attention primarily with Bill Jenkins' performance tests with the Chevy 90-degree V-6 (see the Chevy 90-degree section for Jenkins' discussion of crank, rod, and piston relationships with split-throw cranks, and how they can be "rearranged" on a competition engine). Apparently the situation has been of no major concern to Buick V-6 builders simply because the rod offset doesn't seem to pose any real problems in this engine.

Because of the .120-inch wide flange between adjacent rod journals on the even-fire crank, the centerline of each journal face is consequently moved forward or backward .060-inch. And since rod-journal width was kept the same from odd-fire to even-fire, the

bearing surface on the crank pin had to be extended into the counterweight cheek (which had to be narrowed slightly) on the side opposite the flange. Then, to compensate for the offset of the rod-journal centerline from the cylinder-bore centerline, the big end of the rod is offset .029-inch from the rod-beam centerline, leaving the rod-beam offset .098-inch in the cylinder bore, as well as in the piston (see illustration). Furthermore, the insert bearing shells are offset .009-inch in the even-fire connecting rods, most likely in order to align the bearing with the oil hole in the journal.

Certainly any performance engine builder should be aware of this peculiar alignment of parts in the even-fire Buick. Theoretically such rod-to-piston and rod-to-bearing misalignment could cause bearing, piston, or cylinder wall problems, but such is generally not the case. However, if you are going to rework the crank (or have a billet made), you should consider narrowing the flange—down to as little as .060-inch—to center the rod as much as possible in the cylinder and piston.

Narrowing the flange on a stock crank would require the use of a custom connecting rod with a wider-than-stock big end and bearing—and such modification of the flange would

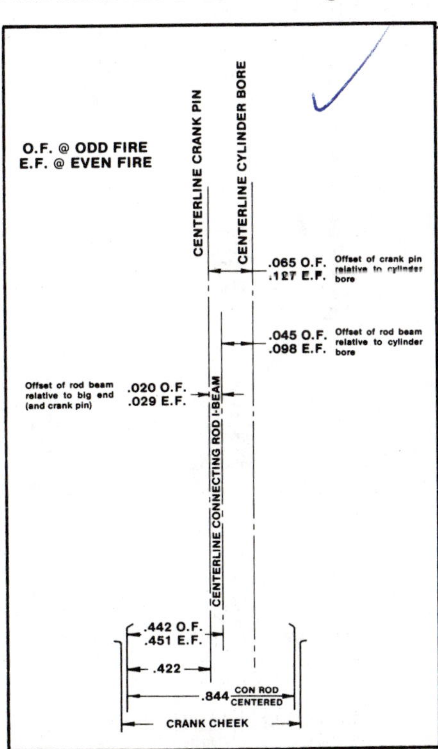

Both the even- and odd-fire crank pins and the connecting rods are offset in relation to the centerline of the cylinder bore. However, according to this diagram obtained from Buick engineering, the offset of the even-fire is approximately double that of the odd-fire.

probably be impossible on a rolled-fillet crank. Best to leave such crank reworking to the very serious racers, who will be machining billets to custom specs anyway. For most performance applications, simply use the turbo crank, have it inspected and prepped, and use top quality bearings such as the fully-grooved set from Kenne-Bell or the TRW (CL77-type) CB1228 P main bearings and MS2960 P rod bearings.

A final point: since the cross-sectional area of pin overlap is critical to the strength of the even-fire crank, never increase the stroke of this engine by turning the rod journals of the stock, cast crank to a smaller diameter. However, destroking the engine by this method does not significantly reduce the overlap cross-section, since material would be ground from the "outsides" of the journals to shorten the stroke. The journals could thus be cut down to Chevy smallblock size (2.10 inches), or to odd-fire size (2.00 inches), to destroke the 231 to approximately 200 inches.

The new 3.0-liter comes with a radically destroked crank (2.66-inch stroke, down from 3.40 inches in other Buicks), giving 181 cubic inches with the standard 3.80-inch bore. This crank necessitates a longer connecting rod (6.33 inches), although rod and main-journal size remain the same. This crank can therefore be swapped into any other Buick block. The drawback to the 3.0-liter crank, however, is the extremely small diameter of the flywheel attaching flange (see clutch section for more details).

CRANKSHAFT BALANCING

Secondly, as discussed in the introduction, the even-fire Buick reciprocating assembly is considerably underbalanced (36.6%) to reduce vertical engine vibration (while proportionately increasing horizontal vibration). Performance builders prefer to balance the V-6 at the standard 50% of reciprocating weight, which will reduce overall unbalance to a minimum. However, to do so means the engine builder must either reduce the weight of the reciprocating assembly (piston, pin, upper half of the rod) by 13.4%—the preferable method—or else add a like amount of "heavy metal" to the crank counterweights. By switching to lightweight forged pistons and lighter wrist pins, this weight reduction can usually be achieved.

If the builder wants to convert the

crank to total internal balance (so that the assembly will not have to be rebalanced every time a flywheel or pressure plate is changed), plus balance the engine at 50%, a considerable amount of expensive Mallory metal will have to be added to the crank.

For engines retaining partial external balance, here's a trick employed by Jim Ruggles to reduce vibration problems in long-distance races—he makes a double-weight torsional dampener by separating a balance ring from one stock dampener and bolting it securely onto the one on the motor. Be sure to align both rings, by matching the timing marks, before drilling them and bolting them together.

An engine that retains stock partially-external balance requires a flywheel and torsional dampener with the proper amount of eccentric, counter-balance weight built in. This amount differs slightly for even- and odd-fire motors. Therefore, stock flywheels or flexplates should never be swapped from one type of motor to the other. Using an odd-fire torsional dampener on an even-fire crank, or vice versa, will also put the timing mark significantly off index.

Most manufacturers of aftermarket flywheels for the Buick V-6 sell the same part for both engines, expecting you to have the whole crank assembly rebalanced (the crank, dampener, and flywheel or flexplate must be balanced as a unit on these engines if external balance is retained). Rebalancing of the entire reciprocating-rotating assembly is of course highly recommended on any performance engine, especially if pistons have been changed. However, if you just want to change the flywheel without tearing the whole motor apart (for instance, to put a lighter flywheel on an odd-fire, stick shift V-6, which would be a quick way to gain some extra rev-ability), some balancers can "match" the balance of your stock flywheel to the new one before you install it.

CONNECTING RODS

The '61 Buick V-6 had the dubious distinction of being the first American production engine to switch from forged to cast connecting rods. In keeping with the spirit of the entire V-6 project, the low-cost cast rod was an economic breakthrough. They're fine for production motors, and are now the standard of the industry, but they aren't as strong as forgings. The original 198/225 cast V-6 rod, readily

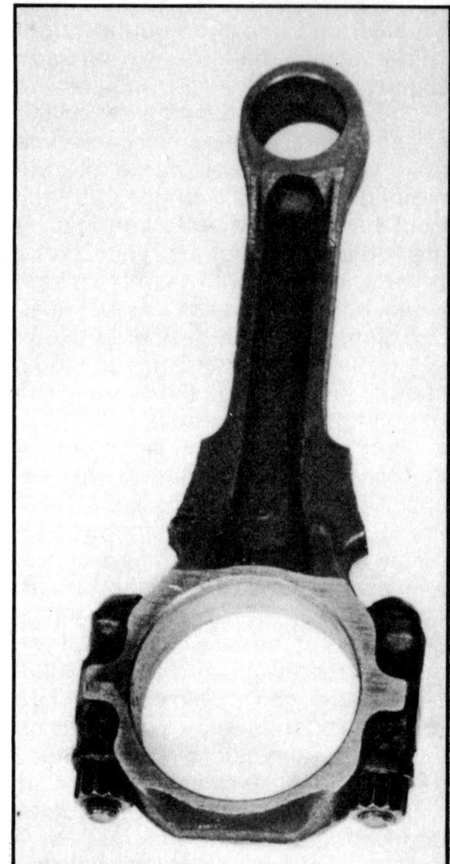
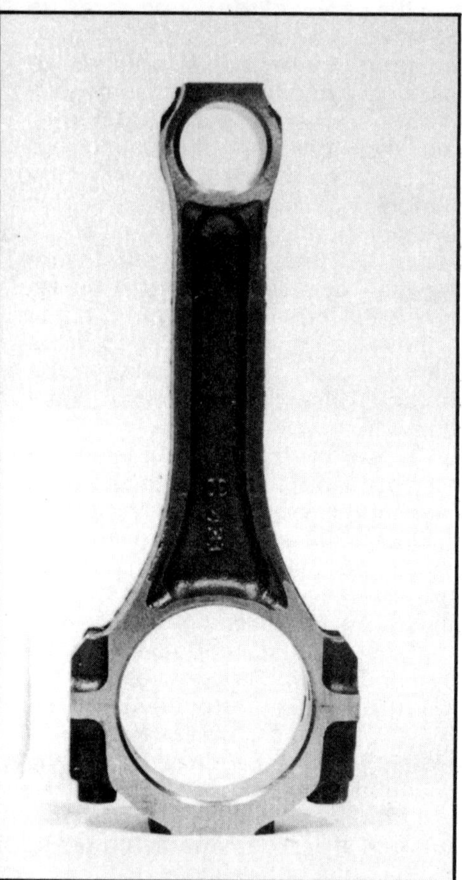

Early odd-fire rod (left) is distinguished by a pair of ears on the beam. The '75-up odd-fire rod (right) is considerably stronger.

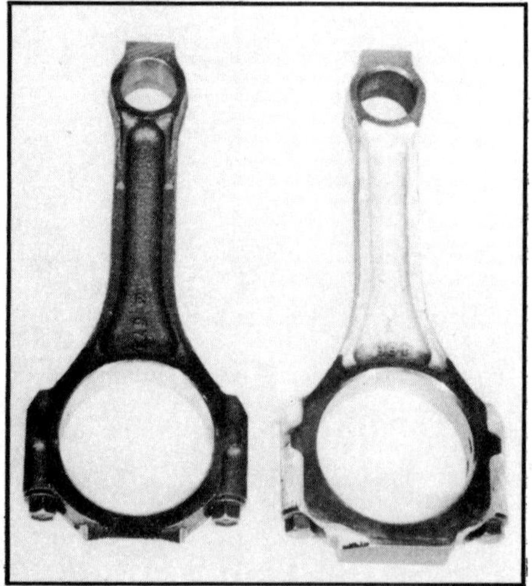

Look closely at the caps on these two even-fire rods. The one on the right, introduced in 1980 production engines, has a considerably thicker cap, though upper portion of rod remains the same.

identifiable by two large "ears" on the shank and through bolts with nuts to retain the cap, is inferior to the '75-up odd-fire production rod. If you are building an odd-fire motor, by all means use the later, stronger, capscrew rods. They are available from the dealer under part number 1249433, or from Kenne-Bell in Magnafluxed, matched sets of six (KB32402).

When Buick switched to the even-fire crank, which has a larger (2.25-inch) connecting rod journal, they continued to use the same connecting rod casting. Since the big-end opening of the even-fire rod has been enlarged, the shoulder cross-section has consequently been reduced, and therefore even-fire rods are not as strong as odd-fire rods.

However, in 1980 Buick switched to a slightly thicker cap for the even-fire rod. The upper portion of the rod remains the same, but the thicker shoulders on the cap will add some strength to the combination. These should be available under service part number 1255667.

Either even-fire rod should be fine for most street performance use. As with any production cast rod, the reliability of the stock Buick piece can be greatly increased by the standard performance prep of Magnafluxing, deburring, beam polishing, and shot peening. Jim Ruggles also strongly recommends X-raying of stock rods to find any "inclusions" in the castings, and to Tufftride them for a considerable increase in strength (after polishing). The stock capscrew rod bolts can be replaced by 3/8-inch SPS bolts, part MS 21250-06014, which are considerably stronger than the stockers.

Although even-fire rods retain the same big end width (cheek to cheek) as the odd-fire, the bearing width on even-fire rods has been reduced to allow clearance for the crank journal fillets. Both sides of even-fire rods are chamfered for fillet clearance as well. Since the bearing diameter differs for the two engines, there should be little problem mixing up components.

As mentioned, the rods in the even-fire engine are also offset in the cylinder bores. As described in the SAE paper describing the even-fire (#770821; *Buick's New Even Firing 90° V-6 Engine,* 1977): "The displacement of the large end of the connecting rod fore and aft in the engine results in an increase in the amount of connecting rod offset in relation to the center line of the cylinder bore and piston... The connecting rod bearing center is within .009-inch of the center of the rod web, however the rod web column is positioned at .098-inch offset from the centerline of the cylinder. The piston pin is centered in the cylinder bore and

the pin bosses in the piston were shortened to provide clearance adjacent to the small end of the connecting rod."

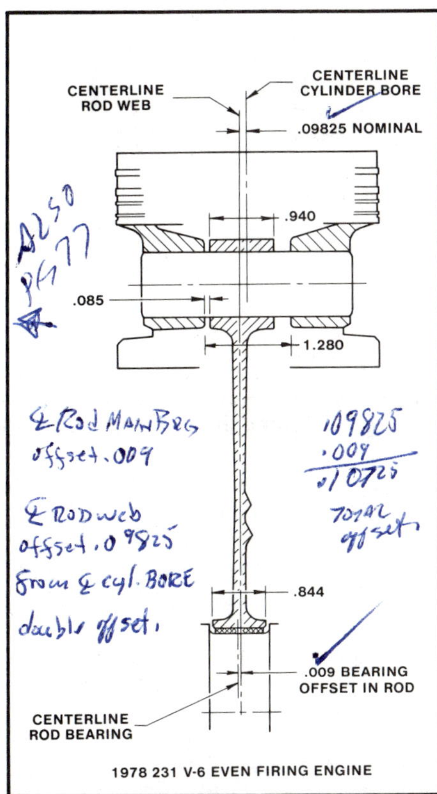

The small end of the even-fire rod is offset nearly .10-inch in the piston, as shown in this factory diagram, plus the bearing is offset .009-inch in the big end.

The small end of the rod remains the same width on odd- and even-fire motors, and it is centered on the beam in both cases, meaning that it is consequently offset in the piston. Obviously, Buick V-6 rods must be installed on the crankshaft in the proper direction for correct alignment with the piston.

Buick has been talking about offering a forged, heavy-duty rod for the V-6 for some time. It is currently scheduled to appear in August 1982 under part number 25500007, and is described as having the same dimensions as the cast rod, except for a .927-inch diameter pin end (large enough to accept a bushing for a full-floating pin) and 3/8-inch aircraft quality through-bolts and nuts. Whether this forged rod materializes on schedule remains to be seen.

However, forged steel rods are available for the Buick V-6 from Carrillo and Crower, in stock (5.960 inches) or longer lengths. They are the currently preferred choice for any sort of long-distance or endurance racing, where class rules allow. Kenne-Bell stocks the Carrillos, plus offers lightweight Howards aluminum rods for the V-6. The latter are applicable to high-revving drag motors; but the thickness of the beams demands close pre-assembly checking to insure there is no interference between the rods and the cam in the V-6 block.

Since the even-fire rod is offset

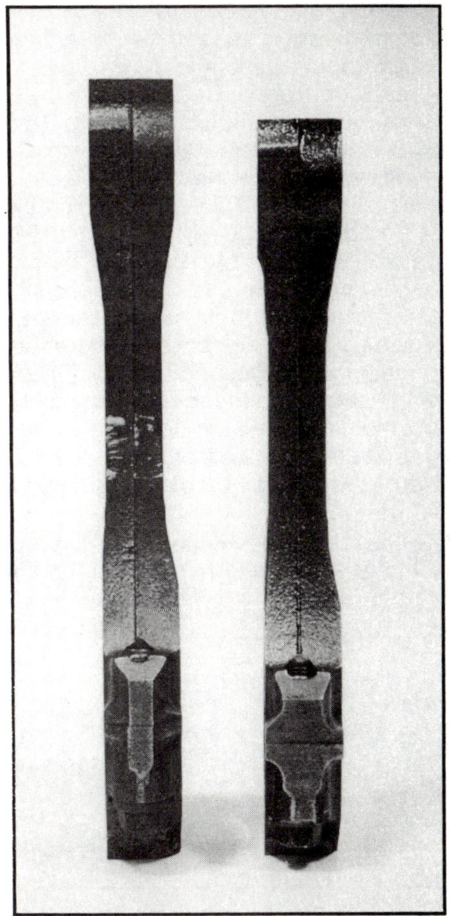

Relative offset of an even-fire rod (left) and an odd-fire rod is demonstrated in this profile. The even-fire rod appears longer because of the larger diameter of the hole in the big end.

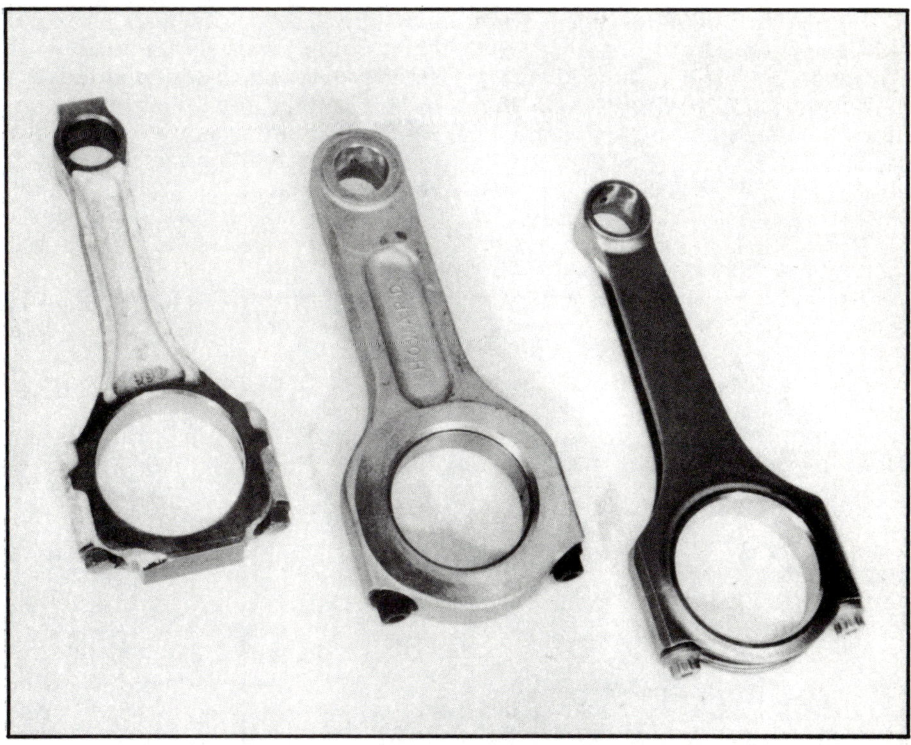

A good variety of custom rods are available for the Buick V-6. Compared to a stock, even-fire rod (left) are a Howard aluminum drag rod and a Carrillo steel H-beam rod.

The stock Buick V-6 piston looks more like an ashtray than a performance part.

Sealed Power and TRW offer excellent, inexpensive forged replacement pistons for the Buick. Shown (left to right) are a Sealed Power stock, replacement-type forged piston (8:1 compression); a Sealed Power forged 9.5:1 piston—excellent for most moderate performance motors; and the same piston with valve reliefs cut in the top for use with cams over .500-inch lift and 280-degree duration.

nearly .10-inch in the piston, some concern must be given to stabilizing the piston on the rod, to keep it from canting in the cylinder. Modifying the rods for full-floating pins would tend to reduce stability. The practice of converting rods and pistons to accept floating pins is aimed specifically at making engine teardown and reassembly easier, and it should be considered only for racing engines where this is a legitimate concern. If you do use floating pins, keep pin-to-bore tolerance at .0015-inch. Otherwise, use pressed pins and note that the pin should be centered in the piston, with the rod offset .098-inch on the pin.

For racing with an even-fire Buick V-6, Kenne-Bell offers .001-inch oversize rod bearing to increase rod clearance to .002-inch without having to polish or otherwise reduce the bearing surface of the crank pin.

PISTONS

Stock Buick V-6's all come with cast, dished-top pistons. Early motors had a compression ratio of 9:1, but this was dropped to 8:1 when the V-6 was reintroduced to the smog era in 1975. That year the bore was also increased .050-inch to 3.800 inches, matching that of the Buick 350 V-8, so the same pistons could be used in both engines. When the V-6 was converted to even-fire in 1977, the same basic piston was retained with one critical difference. To permit the small end of the rod to be offset in the piston, the distance between pin bosses was increased from 1.230 inches to 1.280 inches. This accommodates the rod offset and allows the use of the same pistons on right and left banks. The production even-fire piston was also strengthened in the lower skirt area to withstand greater side loads imparted on it by the offset rod. (Factory turbocharged engines use a slightly different piston than normally-aspirated versions—see turbocharging section).

The new 3.0-liter engine maintains the same bore size as the 3.8-liter, but this short-stroke motor uses a flat-top piston to maintain a moderate compression ratio (8.45:1). Swapping the 3.0-liter piston into the 3.8-liter engine will give a compression increase to approximately 10:1. However, such a swap would not be recommended because this stock cast piston would very likely be damaged by detonation inherent with such a relatively high compression ratio and modern low octane pump gasoline.

Although previous Buick V-6 literature has made little mention of the production piston modification to allow for rod offset in the even-fire motor, this point is critical. Even though stock 231 odd-fire V-6 and 350 Buick V-8 pistons will fit in the even-fire block, they cannot be used in this engine unless the inside faces of the pin bosses are machined to allow the rod to be moved over. Failure to do so would cause the rod to be "cocked" at an angle in the bore, leading to immediate bearing and/or piston failure. Even-fire pistons can be retro-fitted

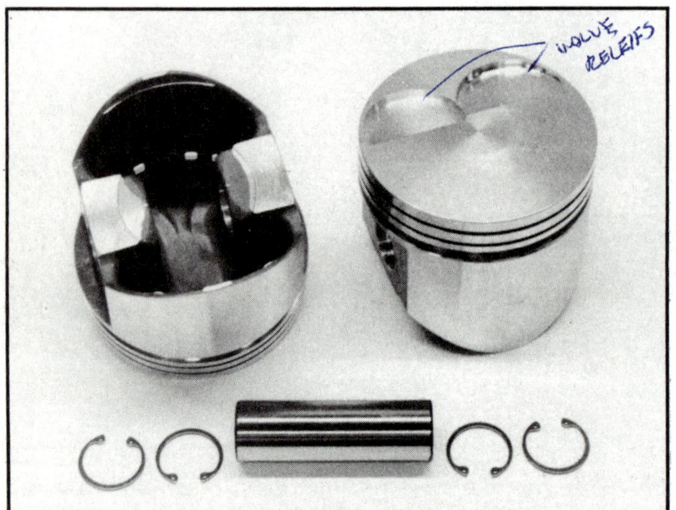

The Diamond 12.5:1 competition piston has been popular with several Buick V-6 builders. It is shown with floating pin retained by dual Spiroloks.

Custom piston manufacturers can make just about anything you might need for a competition Buick—or any other V-6. These examples, among the inventory at Kenne-Bell, include (left to right) a forged blower piston with notched valve reliefs, a 12.5:1 Arias for a 231, and a pro-style 13:1 Arias for use with a longer rod.

into odd-fire motors, but not vice versa. (Consequently many aftermarket replacement pistons are now made to fit the even-fire V-6, but are listed for both engines).

A sure way to check for piston/rod compatibility is to use a feeler gauge to measure clearance between the piston pin boss and the small end of the rod on an assembled short block. An even-fire motor should have .085-inch clearance on the "tight" side, with a larger gap on the other side. Simpler yet, measure the distance between pin bosses on the pistons before installing them. For an even-fire motor, they should measure at least 1.280 inches, face-to-face.

However, finding pistons to fit the even-fire V-6 is no problem since a good variety of aftermarket performance pistons are now made specifically for the 3.8-liter engines. Both TRW and Sealed Power offer forged replacements for the cast production items (TRW L2431F and Sealed Power 2472P for odd-fire; TRW L2452F for '77-'78; and L2471F for '79-'81, even-fire). These pistons have a deeply dished top and are rated at stock 8:1 compression. Although these are not high-performance pistons, per se, they would be a good, stronger substitute for stock pistons in a moderate-boost (5-7 psi) turbocharged or blown street engine.

For normally-aspirated street performance, a good choice would be the Sealed Power 2232P forged, slightly-dished piston, which gives approximately 10:1 compression in the V-6, or the similar TRW L2343F piston, which gives the same ratio. Both of these pistons were designed for the Buick 350 V-8, and may require pin boss machining for use in even-fire motors. Kenne-Bell offers the Sealed Power 2232P piston with machined valve reliefs for use with cams over .500-inch lift and 280-degree duration (KB61003). Plus, they list forged, dished-top pistons (KB62101) featuring a thicker crown and lowered top ring land for use in serious turbocharged or supercharged engines (available in ratios from 6:1 to 10:1).

For all-out racing, the current favorite among Buick V-6 builders is the TRW/Diamond 12.5:1 piston, custom machined from a TRW forging by Diamond Racing (Warren, Michigan). Butch Elkins of Diamond notes that the small Buick combustion chambers allow the use of flat-top pistons—the optimum configuration for even flame propagation in the cylinder—for relatively high-compression ratios, especially in the larger-bore 4.1-liter

engine. This eliminates the need for time-consuming piston dome contouring typical of Chevies and other engines. Elkins also mentioned that turbocharged Buicks like a good quench area in the chamber (provided by a dished piston, rather than a flat-top that provides the same compression with a lower deck height). Further, Elkins stated that he has seen no problems so far in the Buick caused by the rod offset. The Diamond piston is a fully machined flat-top with fly-cut valve reliefs.

TRW markets their own forged, high-performance piston for the Buick V-6 under part number L2469F. It is rated at 11.6:1 (with 48cc chamber volume) and can be fitted with TRW's new ceramic-coated T9112 ring set. Buick is also scheduled to release a 12.25:1 forged flat-top (similar to the Diamond piston) by March of 1982, under part number 25500460. Kenne-Bell markets a full line of "SuperLite" forged racing pistons (KB60003) for the V-6 in ratios from 6:1 to 13:1. (These pistons are custom made by Arias).

Although compression ratios in the range of 12:1 are possible with a flat-top piston in the 4.1-liter V-6 with production heads, Jim Bell states that such figures are not realistic in the 3.8-liter engine, unless the heads are milled or the combustion chambers are welded to reduce the volume. According to Bell, approximately 10.75:1 is the maximum compression attainable with a flat-top in a stock-bore, 3.8-liter with steel shim gaskets (4cc volume) and stock combustion chambers (48cc), with the piston installed at .009-inch deck clearance (which is minimal for performance). The heads can be milled .025- to .030-inch to get chamber volume down to about 43cc, but this will also require deeper valve reliefs in the pistons.

In any case, remember that the actual compression ratio in your engine can only be accurately computed by physically measuring the components (cc-ing the head chamber, the gasket, and the piston in the bore) and dividing the BDC total cylinder and chamber volume by the TDC total chamber volume. "Rated" compression ratios listed by piston manufacturers can only be viewed as rough approximations. The "actual" compression ratio in a specific engine assembly will usually vary from these approximations.

TRW engineer Dennis Novotny cautions that production Buick V-6's used a variety of piston weights in different years. To avoid upsetting the

balance in your motor when changing pistons, it is a good idea to weigh the stock combination and replace it with a piston and pin of identical weight (if possible). If the replacement piston weight is significantly different, you might be able to swap wrist pins to bring them in line (production engines used pins of 108, 153, and 162 grams).

Of course any performance engine should be completely and accurately rebalanced, especially if pistons are being changed. It would be easiest to rebalance the even-fire motor at the production 36.6%. Rebalancing a high-performance even-fire at 50% of reciprocating weight will require the use of an extremely light piston and pin or the addition of Mallory metal inserts to the counterweights on the crank (or both).

Finally, it is common practice for builders of racing engines to convert connecting rods for floating piston pins. This is done largely to facilitate rapid engine teardown and reassembly. However, floating pins have their drawbacks. Drilling and bushing stock-type rods to allow the piston pins to float weakens the small end of the rod (material thickness around the pin bore is reduced), and Spiroloks or other retainers used to keep the pin in the piston have been known to pop out, causing piston or cylinder wall damage. Pressed pins, though tedious to install, are much more bulletproof, and should always be used in street-type motors which are not subjected to constant teardowns. In the case of the even-fire Buick V-6, where the rod is offset in the piston, the pressed pin will also help to stabilize the piston on the rod and keep the piston laterally perpendicular in the bore.

CYLINDER HEADS

From 1961 through 1976 the Buick V-6 used the same cylinder head, a design inherited directly from the small-bore, aluminum V-8. It is nothing to get excited about. Valve sizes were only 1.5 inches in the intake, and 1.3125 inches in the exhaust, with correspondingly diminutive ports. Further, these heads were quite heavy, having been taken from tooling for aluminum parts. They are readily recognizable by a raised outer valve-cover rail and traditional Buick "nail-head" covers, which are nearly vertical. These early heads feature shaft-mounted rockerarms (like all Buick V-6's) with the shafts supported by three detachable aluminum pedestals. Pre-'75 heads have cast aluminum non-adjustable rockerarms which oil

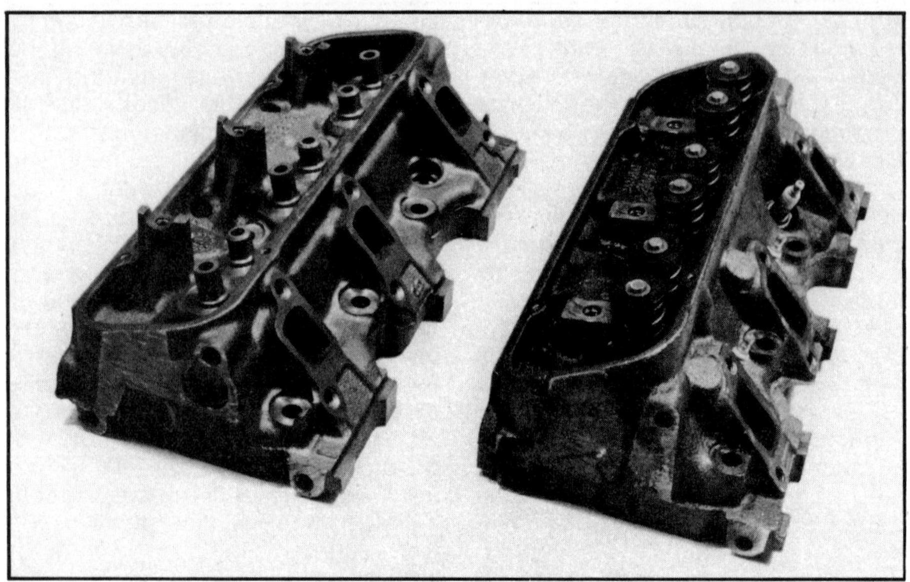

Comparison of pre-'77 cylinder head (right) with '79-up "high-port" reveals several differences beyond obviously elevated exhaust ports. Earlier head has raised valve cover rail on exhaust side, and doesn't have cast-in pedestals for rockershafts. Also note freeze plug in front of later head.

through the rockershafts. All 1975 and later heads have stamped steel rockers, held in place on the shafts by press-in teflon buttons, and they oil through the pushrods.

The early heads are strictly low-performance items, and require a considerable amount of reworking to produce even reasonable horsepower. Unfortunately, '77 and later heads cannot be used on '76 and earlier blocks (without modification) because of a change in water passage shape and location (see block section). If you are planning to build a high-performance Buick V-6, you would be wise to start with a '77 or later block so you can use the later high-port head. If you are building a pre-'77 motor, however,

Kenne-Bell still offers an extensively ported "Stage 3" head (KB12002) with larger 1.71-inch intake and 1.5-inch exhaust valves to fit this block. They claim these heads are good for 25-30 horsepower and flow almost as well as the high-ports up to 6000rpm, with the high-port heads showing a decided advantage beyond this point.

In model year 1977 Buick revised the head casting, primarily to save weight. Immediately obvious is a lowered valve cover rail and a switch to the current "flat-type" valve cover. These and subsequent heads can also be identified by a freeze plug in each end. Other changes included a switch to cast-in rockershaft pedestals, revised water passages, narrowing of the face,

relocation of the spark plug farther away from the valves (for smog control), and an increase in valve sizes to 1.63 inches in the intake and 1.42 inches in the exhaust.

The exhaust crossover passages were also enlarged on these heads, which means that manifolds designed for the earlier heads will not work (because they do not cover the heat passages), even though port size and bolt configuration remained the same. Another minor change was that pushrod holes in these (and subsequent) heads were enlarged so that the pushrods could be slightly angled to allow the use of identical rockers on both intake and exhaust. Prior heads use "right" and "left" rockerarms with offset pushrod pockets. These new heads were used in model years '77 and '78 only, and from a high-performance viewpoint they still weren't anything to shout about.

The breakthrough in head design for the Buick V-6's came in 1979, as a direct result of extensive testing and modification of the engine for Buick by Smokey Yunick. The new heads, installed on all production V-6's from that date on, feature significantly taller intake and raised exhaust ports, as well as valve sizes increased to 1.71-inch intakes and 1.5-inch exhausts. These heads, *in ported form*, showed impressive 43% gains on the intake and 83% on the exhaust on Smokey's flow bench over the stock '77-78 heads. They are definitely the parts to use for a high-performance V-6.

Simply bolting these heads onto the '77-78 blocks will greatly increase performance, and by now you should be able to pick them up relatively inexpensively in the junkyards. In fact, by welding or plugging and then re-drilling the enlarged water holes in these heads (using a pre-'77 head gasket as a template), you could probably make them work on earlier blocks. Obviously, you will have to use an intake manifold compatible with the taller intake ports, and you should also switch to compatible exhaust manifolds (or headers), since exhaust port shape is slightly different, as well as raised, although the bolt pattern remains the same.

Even though the high-port heads yield impressive gains in bolt-on form, according to Smokey Yunick they can be significantly perked up with some minor porting and polishing. The final increment of his "5-step" plan for the Buick is what he calls a "dollar port job" on these heads. Specifically, this quick porting consists of blending in the sharp edges immediately beneath the

Early Buick heads used cast aluminum rockerarms on a shaft supported by removable aluminum pedestals. All '75 and later heads have stamped, steel rockers with shaft pedestals cast in the head.

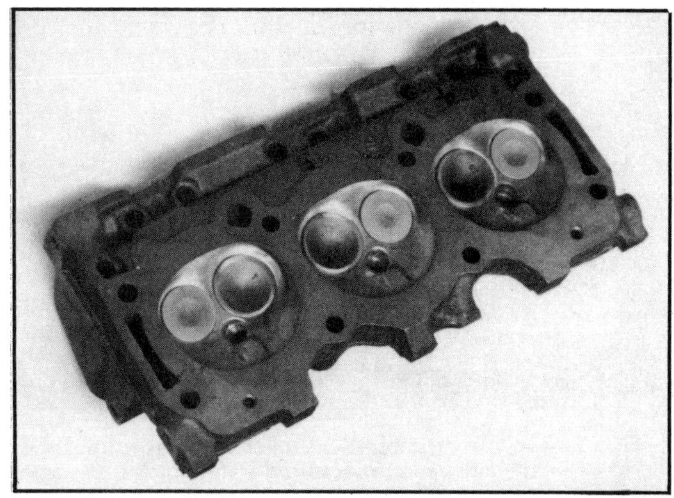

All Buick heads have open combustion chambers. They also have relatively small valves, a carryover from the small-bore 215 V-8 on which the engine was patterned. This pre-'77 head has 1.625-inch intakes and 1.425-inch exhausts, plus a centrally-located spark plug.

This post-'77 head has been given a full "Stage-3" porting and polishing by Kenne-Bell with valve pockets enlarged for 1.71-inch intake and 1.50-inch exhaust valves—obviously about as big as you can go without moving the valveguides. Also note the factory relocation of the spark plug, supposedly to reduce NOX emissions.

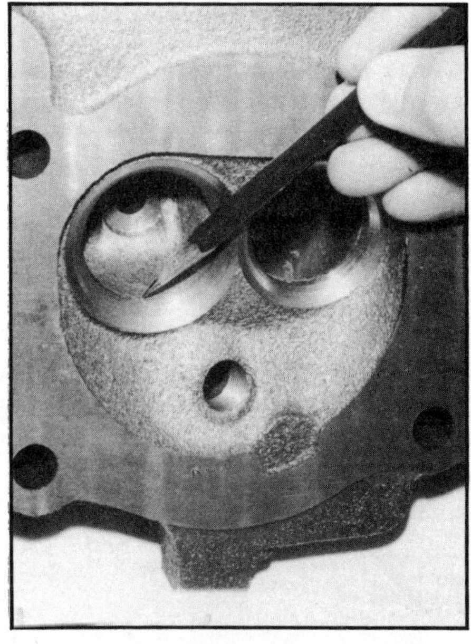

Smokey's "dollar port job" consists of back cutting the first 1/4- to 1/2-inch of the port below the valve seat, as shown, followed by a quick clean up grinding of the port runner to remove excess casting flash.

The new Buick heavy-duty Stage I cylinder head is very similar to the 4.1-liter production high-port design, but features internal casting changes to strengthen the deck and allow greater freedom to grind the ports larger.

valve seats and then polishing the remainder of the ports with hard rolls on a hand grinder. On Smokey's dyno this minor cleanup of the stock castings was good for 28 extra horsepower. Whether the average home builder can achieve the same results in "only a few minutes" with the grinder is doubtful, however, since Smokey spent plenty of time practicing and testing to find out which licks were best. But there is no doubt that these heads can be improved with the cleanup procedures described.

Butch Elkins, the "head master" at Diamond Racing, was more specific about modifications to the production high-port heads. He notes that the exhaust ports are excellent (the exhausts were quite deficient in the early heads), with potential flow better than the best a Chevy has to offer. However, the intake port needs work, specifically in the short-turn radius (what he calls the "pressure ridge") over the nose of the port. "We had to remove enough material that a lot of times we created sprinkling systems," said Elkins. "In order to get anything out of it, we *had* to get that close to water." Typical core shift in the heads makes porting this area slightly experimental. He also notes that the intake port floor is a little too flat on the production casting.

To remedy these problems, Buick is introducing a new heavy-duty head, known as the "Stage I" (part 25500009), supposedly to be available over-the-counter as of February 1982. Basically a refinement of the production high-port head, this casting will feature revised water jacket coring to give more meat under the critical intake port area, plus it adds some material to the intake port floor. This head also has a reinforced and enlarged deck to match that on the new 4.1-liter, heavy-duty block, although it should retrofit onto all '77 and later blocks. In fact, Buick claims that it should fit on any Buick V-6 block, but be sure to check that no water leaks will be created by mismatched casting faces or water passages. One easy way to check compatibility would be to match a head gasket for the block in question to the heavy-duty head.

The exhaust heat crossover passages have also been eliminated on this head, and at each end of the deck surface two smaller water holes have been substituted for the larger, elongated water passages used on previous heads.

Future plans at Buick call for a further revision of this design, tentatively labeled the "Stage II" head. No release date or part number has yet been assigned, but the development program is calling for wider spacing between the valve guides to allow 2.02-inch intakes and 1.6-inch exhausts, further port reshaping, plus a provision for adding two more head bolts around each cylinder. These heads could materialize within a couple of months of the Stage I release...but as of this writing they are strictly in the "dream" stage. The heavy-duty heads will be cast in iron, and Buick states that they have no plans in the near future to produce an aluminum version.

HEAD GASKETS AND COOLING SYSTEM

The fact that the Buick V-6 has only eight head bolts, plus the fact that the deck surface on production blocks is relatively thin and the core holes and water passages are large, led to early concern over adequate head sealing on this engine, especially with in-

Note the extended deck on the intake side of the combustion chambers on the Stage I head. The two distinctive "D-shaped" holes and the bridged water holes at the front and the rear of the head can also be seen here.

creased cylinder pressures and operating temperatures. Of further concern is the close proximity of adjacent cylinders, which leaves only .415-inch gasket width between bores on the 3.8-liter, and only .240-inch on the 4.1-liter block. Remember, the block is based on a design originally intended for 2.80-inch bores, and center-to-center cylinder spacing has remained the same. Production engines come

A good selection of custom valve covers exists for the Buick, including offerings from Kenne-Bell, Holley, Offenhauser, and Moroso. The Kenne-Bell cover has clearance for roller rockers and fits early or late heads. Most other types fit only '77-up heads.

with thin (.020-inch) embossed steel head gaskets, which are very durable and heat resistant, but which do not offer much "crush" or resiliency to contend with block or head warpage.

The aluminum V-8 from which the V-6 was adapted had the advantage of an extra row of head bolts along the outboard side of the cylinders. The Oldsmobile motor, as opposed to the Buick V-8, had another row of bolts along the intake side of the head as well. These bolts went through the rockershaft pedestals, serving to attach the shafts, pedestals, and heads. The V-6 heads and blocks could probably be modified to accept this extra row of head bolts through the rocker-

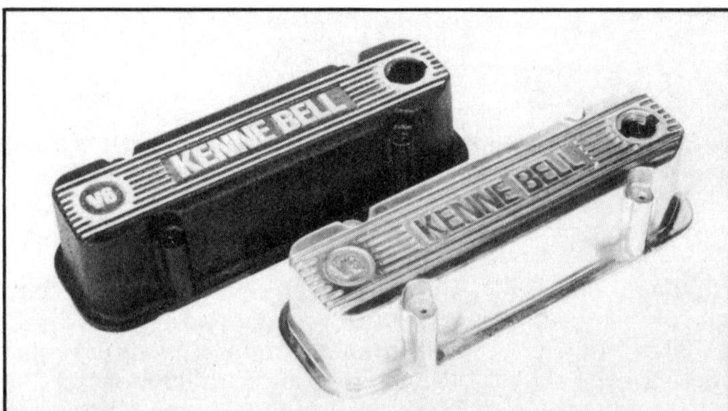

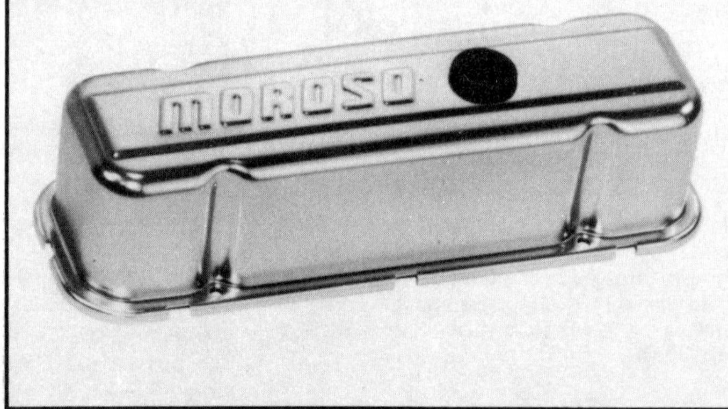

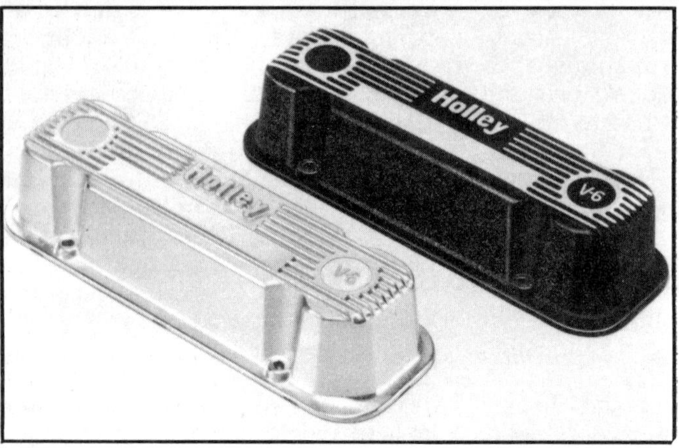

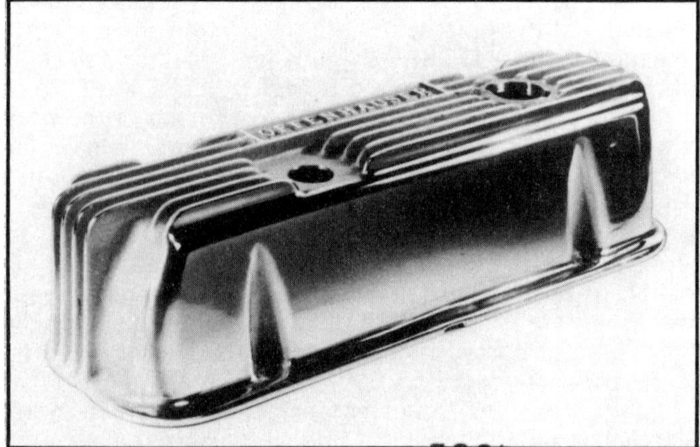

shaft stands, but it would be quite a bit of work, and probably not worth the results. Hopefully the proposed Stage II, heavy-duty heads will incorporate this feature, along with provision for an outboard row of head bolts, greatly improving the head-to-deck clamping force.

Heavy-duty blocks are cast with a thicker (.50-inch) deck and have smaller water holes than production versions, which greatly improve deck strength and stability. The early heavy-duty 3.8 block also had blind head-bolt holes to further improve strength, but these caused some problems with water circulation, and are not a feature on current heavy-duty blocks. The new 4.1-liter, heavy-duty block also has the top portions of the cylinder walls cast together (siamesed) to further strengthen the deck and minimize cylinder distortion (later heavy-duty 3.8 blocks may be the same).

When Smokey Yunick began testing of the Buick V-6, two of his initial concerns were head sealing and coolant flow. Even with a prototype heavy-duty block, he felt the large water passages (primarily the long "dogbone" holes at the front and back and the D-shaped holes above the cylinders) needed to be welded shut to help tie the cylinders to the blockcase, and then drilled with smaller holes to restrict coolant flow. It should be pointed out that these large passages, especially the ones in the heads, are put there primarily to allow removal of coring sand after casting, rather than to facilitate water flow. You will notice that some of these large cavities are covered by much smaller holes, or completely blocked off, by the stock head gaskets.

Secondly, Smokey felt that coolant flow through the heads was both excessive (heat is energy; you don't want the water to carry *too* much of it away) and uneven. His solution, besides restricting some passages, was to route water out of the front of each head into a Y-pipe rather than through the intake manifold to the stock water neck. This system has been used on other V-type motors (i.e., Chevies) for circle-track or other long-distance racing, and its main objective is to help balance the coolant flow through each bank (cylinder head) of the engine, as well as to improve flow at the bottom of the head jackets, primarily around exhaust valves.

To attach this Y-pipe water manifold Smokey drilled and tapped a hole in the front of each head, directly below the freeze plug and about .50-inch above the head face. The Y-pipe is made from regular .75-inch (o.d.) plumbing pipe and elbows, with each leg of the Y-pipe being about 12 inches long, before being joined together, forming a single tube to accept a radiator hose (to carry coolant to the top of the radiator). You could easily fabricate such a water manifold yourself, but it is also available from Moroso in kit form (part 6380—for Chevies, but adaptable to the Buick).

A different, slightly more complex approach to cooling improvement in the Buick V-6 was devised by Doug Roe and is now used by many engine builders (it has consequently been shown in several Buick V-6 articles, but never fully explained). Here is the way Roe's system works: in the stock Buick (as in most modern engines), cooled water from the radiator is pumped into the block by the water pump, flows around the cylinders, then enters the heads primarily through the large holes at the back. The water then flows forward through the heads, out at the front, into the water passage in the intake manifold, past the thermostat, then through the water neck and upper hose back to the radiator. Thus the majority of the water is preheated in the block before reaching the head, and the uni-directional flow through the head (from back to front) does not necessarily circulate coolant sufficiently around the combustion chambers, especially the exhaust valve

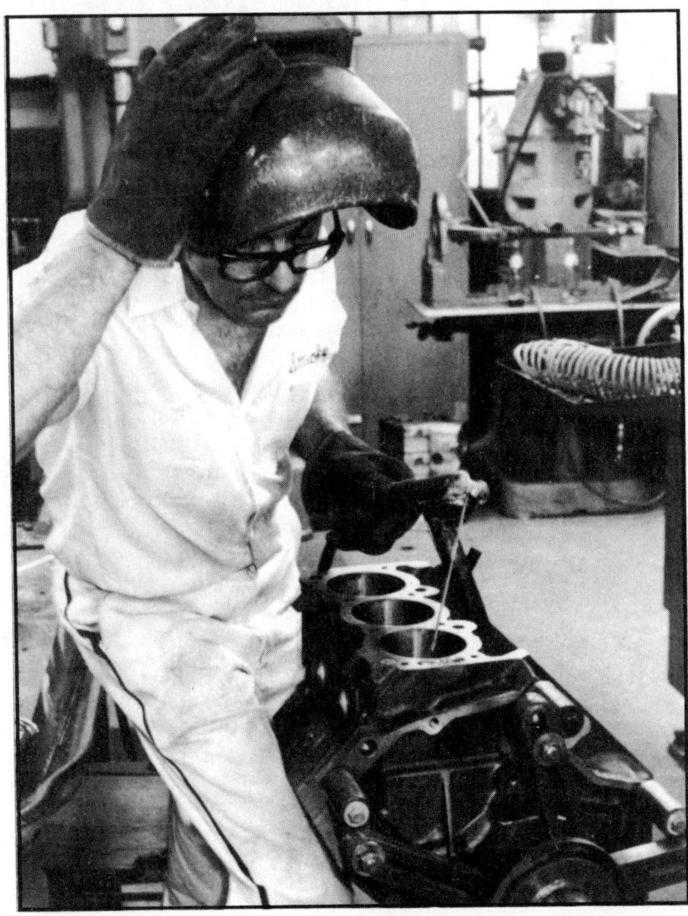

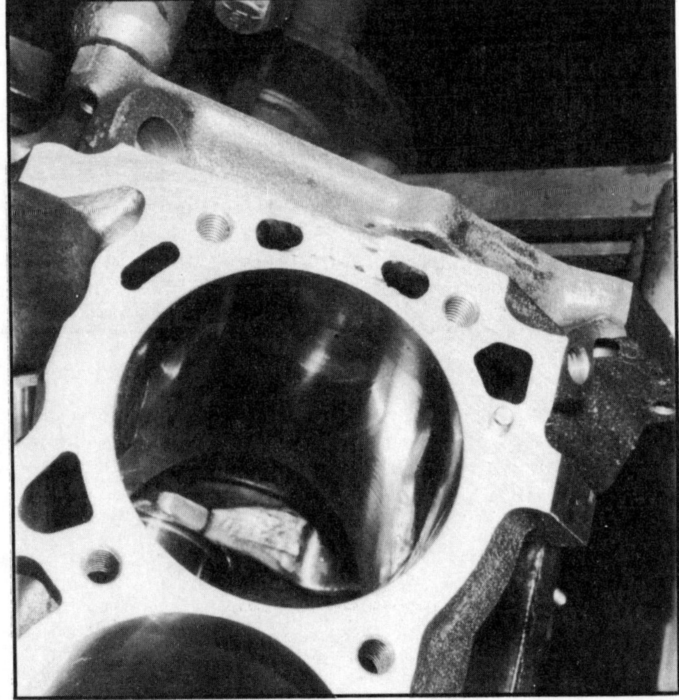

Smokey Yunick feels it is necessary to weld up the middle of the "dogbone" water hole at the rear of the production deck to "tie the cylinder to the block." This step naturally requires resurfacing of the decks. Other builders consider strengthening unnecessary and partially fill the hole with epoxy or screw-in plugs to restrict water flow. Either step will be unnecessary on the new heavy-duty block.

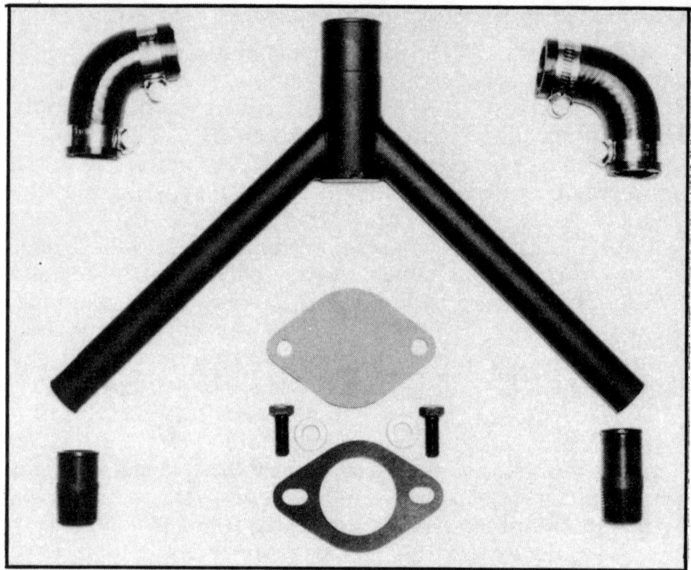

The Smokey "Y-tube" water neck is available as a kit from Moroso, and is shown here installed on Ray Baker's V-6 stock car.

seats.

As Roe explains, "Flowing water takes the path of least resistance, which is often straight through across-the top of the head and out, without cooling the chambers." Therefore Roe restricts the large water passage at the back of the block and head, and drills new holes around each cylinder, to force the water to "back up" in the block and "percolate" evenly up around each cylinder and combustion chamber.

It's a relatively simple system in concept, but it does require fairly precise sizing and placement of the new water holes to make it work smoothly and accurately. Whereas Smokey Yunick welds up the long holes at the front and rear of the block and head to add strength, and Jim Ruggles uses screw-in taper plugs (both of which require re-machining of the surfaces), Roe simply fills them with epoxy or Devcon since he feels that strengthening is not necessary here. The front passage is closed off entirely, and the rear passage is restricted to two 3/8-inch diameter holes, located at the outer ends of the "dogbone." Roe also adds a .25-inch hole in the lower rear corner of the head gasket (above an existing .50-inch hole in the block), and drills a corresponding .25-inch hole in the head which must be angled (at about 45 degrees) towards the rear and center of the head to reach the water jacket. Likewise, the small holes in the head gasket covering the larger D-shaped holes above the cylinders in the heads are enlarged to .25-inch. The most important change, however, is to add two .25-inch holes between each pair of cylinders in the block, heads, and gaskets. These holes are spaced approximately .25-inch in from the head bolt holes.

This design is similar to water passages standard on Chevy small-block V-8's, and allows coolant to trickle up around each cylinder and combustion chamber. To help reduce water leaks and to match passages on various blocks and heads, some builders also fill in the large D-shaped holes above the cylinders and drill matching .25-inch holes in the heads and block. On heavy-duty 3.8 blocks with blind head-bolt holes, Roe also found it necessary to drill a pair of holes, on an angle through the D-shaped ports and towards each other, to allow steam to escape from a pocket formed between the top inner two head-bolt bosses in the block. The hole drilled through the front D-shaped hole should be 1/8-inch, the one through the rear hole 3/16-inch.

Obviously trying to describe these block and head modifications in great detail is difficult (even photos don't help much). As Roe says, "If you have a good understanding of coolant flow principles in general and specific water-jacket coring in the Buick, common sense will tell you where the holes should go and what size they should be. *Otherwise, let someone else more experienced do it.*" Jim Ruggles, who has been building quite a few Buick V-6's for IMSA racers as well as for Buick test cars, drilled his block torque plate so that it doubles as a drilling fixture to add the extra water holes quickly when the block is honed.

We must strongly emphasize that the relatively complicated water system reworking outlined above is necessary *only* for engines which will be used in long-distance, extreme-performance situations such as road-racing, circle-track, or off-road com-

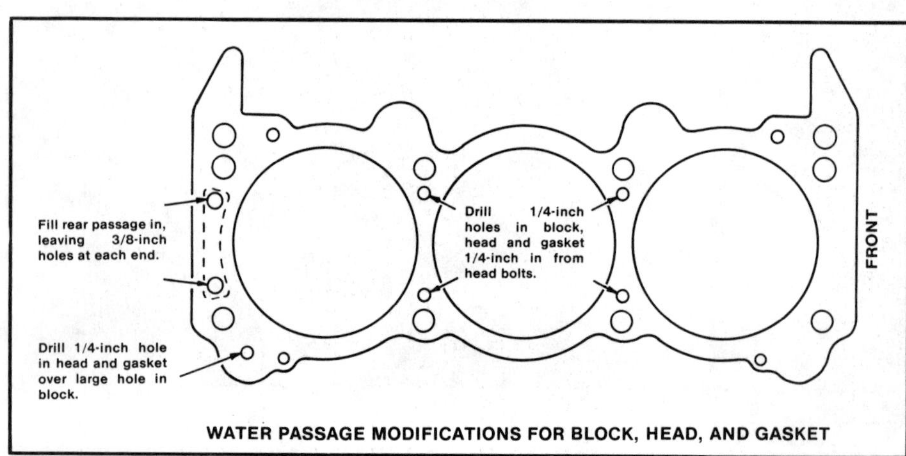

This diagram illustrates cooling system alterations devised by Doug Roe for the Buick V-6. The most important changes are the restriction of the large "dogbone" hole at the rear of the deck and the drilling of new holes between cylinders in the deck, gasket and head.

petition. Street motors should not need these modifications, and they would be a total waste of time on a drag-race engine.

Since drag racers often control engine temperature by draining and refilling the cooling system between rounds, they will probably want to add one or two petcocks to the Buick block, which comes with no provision for a water drain. You can drill and tap the side of the block to accept a standard drain valve, but do not drill in an area below the bottom edge of the two freeze plugs on either side or you will hit (crankcase) oil instead of water.

Although some initial concern was expressed over head-gasket sealing on the Buick V-6 and some problems were encountered with the first heavy-duty 4.1-liter blocks used for the Indy pace car program, most builders report little actual problem with head gaskets on this engine. All Buick builders agree emphatically, however, that head studs, instead of bolts, should be used for serious racing. Studs are available from a variety of sources, and complete stud kits for the Buick are sold by Kenne-Bell (KB32001).

HEAD GASKETS

The stock steel shim head gaskets seem to work well in most cases. Several builders, however, prefer to stack two of these gaskets together for better sealing (which compress to a

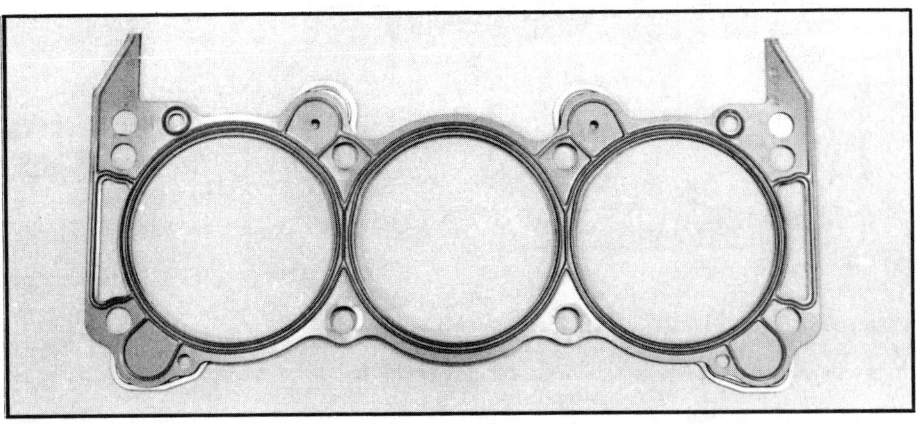

Minor head changes are difficult to spot on Buicks unless you closely compare head gaskets or match gaskets to blocks and heads. Here a pre-'77 gasket is placed on top of a '77-79 gasket, and you can see that one is larger in certain areas. Matching the wrong head to a block or selecting an improper head gasket can result in water leaks.

total thickness of .039- to .040-inch). Other builders prefer the new Fel-Pro Perma-Torque composition gasket, which has a folded steel ring around each cylinder. These are available for the '77-up 3.8 block under part number 8142 PT-1 and are approximately .040-inch thick.

If cylinder sealing were actually a chronic problem of the Buick V-6, builders would have resorted to O-ringing the heads long ago, but such is not the case. As a matter of fact, cases of water seepage between head and block seem to be more common than blown head gaskets. After performing water passage modifications on an engine, Doug Roe has noticed that water will "weep" up the bolt holes in the heads. To cure this problem he cuts a slight chamfer around the bolt holes in the head face, then rolls a thin rubber O-ring down over each head stud after the gasket is in place to seal the hole. In other cases, it might help to add a bit of sealant, such as high-temp RTV, around each water passage, or else coat both sides of the head gasket with a sealer like Permatex High Tack.

The preceding comments refer primarily to the 3.8-liter engine. The new 4.1-liter engine presents other problems. These motors come with a gray composition Perma-Torque gasket which seems to be fine for street use. Under heavy racing conditions, however, especially on blocks which have been slightly overbored (leaving as

Water leakage problems result from the peculiar shape of the deck or sealing surface on Buick heads and blocks. A comparison of a pre-'77 head (left) with a '77-78 head shows areas in which size and shape of sealing surfaces vary.

Comparison of a '79 head (top) with an '80 4.1-liter head shows considerably different sealing surface area on the intake side of the combustion chambers. Block and gasket surfaces must match the heads.

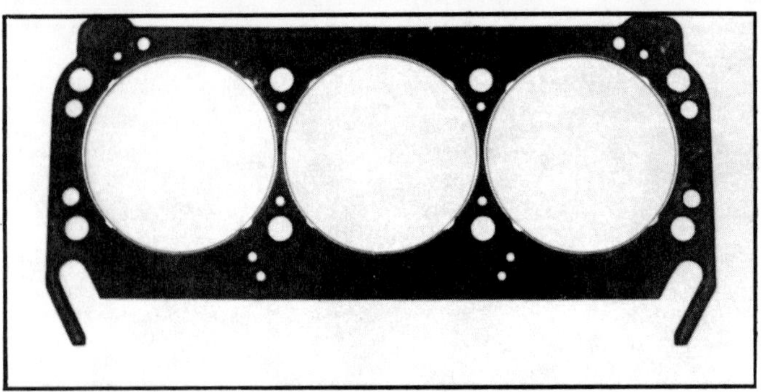

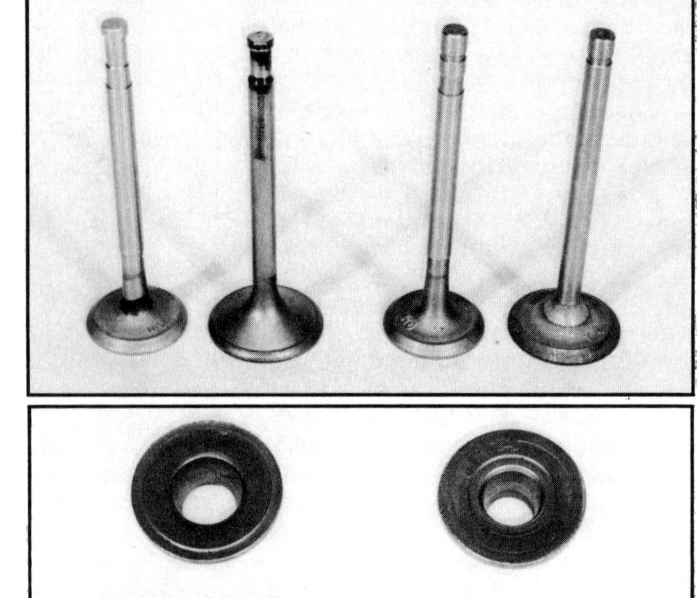

This is a prototype of a new "universal" gasket in the works at Kenne-Bell. Similar to a heavy-duty 4.1-liter gasket being developed by Buick and Fel-Pro, it incorporates stainless steel O-rings for cylinder sealing. In addition, the K-B gasket includes extra holes which can be used as a drilling template for deck and head water hole alterations. Also note cutouts at lower left and right corners for better oil drainback from heads to pan—an alteration you can make to any Buick gasket. When it is finalized, this gasket will fit 3.8- and 4.1-liter engines.

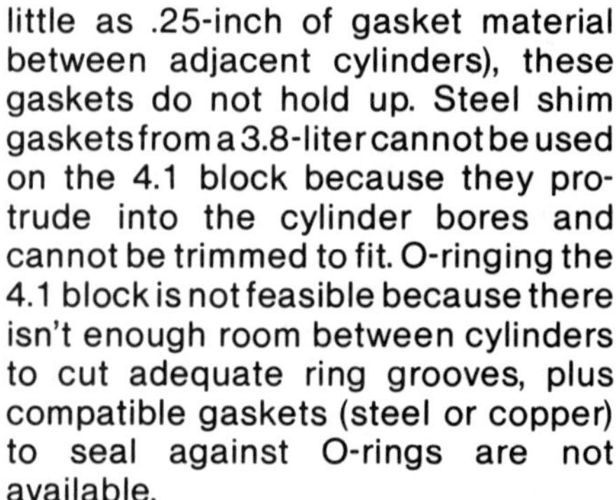

Buick valves are small and anything but streamlined. Pre-'77 valves (left) use a different split lock and retainer (shown on left on lower photo), while '77-78 valves switched to Chevy-style keepers and retainers.

little as .25-inch of gasket material between adjacent cylinders), these gaskets do not hold up. Steel shim gaskets from a 3.8-liter cannot be used on the 4.1 block because they protrude into the cylinder bores and cannot be trimmed to fit. O-ringing the 4.1 block is not feasible because there isn't enough room between cylinders to cut adequate ring grooves, plus compatible gaskets (steel or copper) to seal against O-rings are not available.

During the '81 Indy pace car project Buick worked with Fel-Pro to design a new composition gasket for the 4.1, similar to the Perma-Torque, but incorporating a stainless steel ring (much like a built-in O-ring) around each cylinder. Prototypes of these gaskets seemed to solve the head sealing problems on the heavy-duty 4.1-liter engine in the pace car. However, as of this writing, these gaskets have not been released to the public. The delay seems to be over exact placement of water passage holes in the final design, but it appears that the new gasket will be made to fit the heavy-duty 4.1 block and the heavy-duty head, and will be released after these parts become available. Whether this gasket (or a different version of it) will be made to fit 3.8-liter block and production heads remains to be seen. However, if water passages can be made to work, this new O-ring gasket could very well be used on the 3.8 engines.

The fact that Buick has made several minor changes to water passages and deck configurations on the V-6 over the years plays havoc with builders trying to match certain heads to certain blocks. Jim Bell has been working with McCord for some time to devise a "universal" Buick V-6 head gasket that will allow interchange of '77-up heads on pre-'77 blocks and vice versa without modification to water passages. However further changes in 1980 production heads, plus introduction of 4.1-liter engines and the new heavy-duty parts has sent them back to the drawing board.

They now plan to have this gasket design finalized after the introduction of the heavy-duty block and head (though it has already been issued part number KB18009). This gasket should be of a composition design incorporating a stainless steel O-ring like the Fel-Pro, and will be made to fit either 4.1- or 3.8-liter motors. If and when it does become available, it will greatly simplify the problem of cylinder head interchange on Buick V-6's and will allow the use of the superior '79-up, production, high-port heads on pre-'77 engines.

VALVES AND ROCKERARMS

Valve head diameters on all production and even the proposed Stage I heavy-duty Buick heads are relatively small. Maximum valve size is limited to 1.80-inch intakes and 1.50-inch exhausts by the location of the valve guides. By offset drilling and bushing the guides to space them .100-inch further apart, Jim Ruggles has in-stalled 1.937-inch intakes and 1.550-inch exhausts in these heads. To install larger valves than this, the guides would also have to be moved closer to the center of the cylinder (these changes may appear on the proposed Stage II heads).

However, size is not the only drawback to stock Buick valves. They are also rather heavy and poorly shaped under the head for racing purposes. Consequently most builders in the past have substituted either Chevy or Ford high-performance valves (which have the same stem diameter as the Buicks), cutting them down to proper head size. A further advantage of the Chevy or Ford valves is that they have slightly longer stems than the Buicks, which allows the use of taller or thicker-wound valve springs for increased pressure, plus allows for increased opening of the valves before the retainers hit the valve guides. Both are problems on the stock Buick heads.

Smokey Yunick and other builders have used Chevy high-performance 2.02- and 1.60-inch valves, which are available from numerous manufacturers (including Chevrolet), cut down in diameter to fit the Buick heads. These valves are .065- to .070-inch longer than the Buicks, which gives proper rocker-to-valvestem contact geometry with cams of over .500-inch lift at the valve. With bigger cams (around .650-inch net lift) Smokey uses the .100-inch longer Chevy valves available from some aftermarket suppliers. Jim Ruggles prefers 289 Ford high-

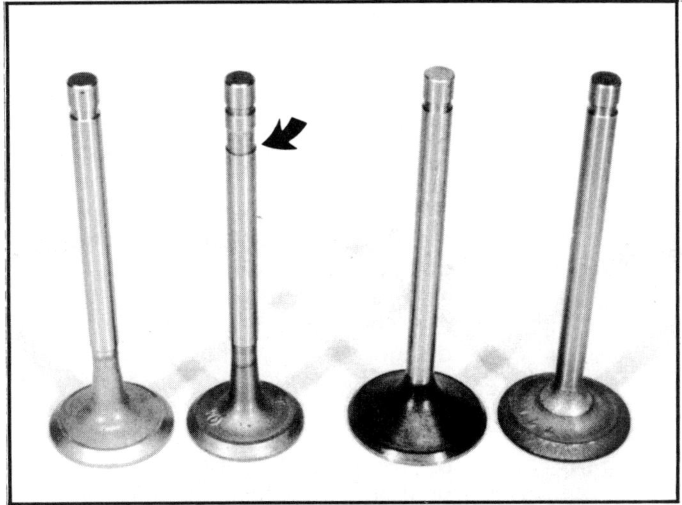

The '79-up production valves (right) are significantly larger. The intakes measure 1.710-inch and the exhausts 1.50-inch. They can be back-fitted into earlier heads if the pockets and seats are enlarged. The newer valves also eliminate the stepped shoulder on earlier exhaust valves (arrow), which can tear up performance-type valvestem seals.

The slightly longer length of Chevy performance valves allows the use of longer valve springs and provides better rockerarm geometry with high-lift cams. However, the head diameter of the valves must be cut down to fit the Buick valve pockets. The Chevy valves are obviously lighter and better shaped for performance.

performance valves, which come in 1.900- and 1.600-inch diameters and have slightly shorter stem length than the Chevies.

In either case, using the longer valves requires that the rockershaft be raised a proportional amount to maintain proper geometry, and longer pushrods must be used. To raise the shaft on the cast-in pedestals, you will have to fabricate crescent-shaped shims to fit between the shaft and the pedestal—not a simple task. Baker Engineering offers Buick rockershaft shims in .185-inch thickness; they may also be available from aftermarket cam companies. One alternative is to machine the pedestals flat to the base of the "cup" and then cut down pre-'77 aluminum rockershaft pedestals to the correct height, and bolt the assembly in place.

Obviously the above modifications require considerable time, skill, and money. They are for serious racing, where such extra effort is worth the expense. The problem is that there have been few alternatives to hybrid valves for the budget racer. Kenne-Bell offers pre-modified long stem valves in 1.75-inch intake and 1.5-inch exhaust sizes, either in steel or titanium, but don't expect them to be cheap. They also catalog oversize (1.71-inch, 1.50-inch) standard-type valves to retrofit in earlier heads, either with pre-'77 (long) valve lock grooves or with current (Chevy) style.

The good news for racers and street engine builders is that TRW will soon (Summer 1982) release performance valves specifically for the Buick, part V-3124X for the intake and part S-3123

for the exhaust. They will have considerably improved head design (radius-flow, tulip shape), be made of swirl-polished stainless steel, and come in standard Buick valve length so you won't have to shim the rockershafts, change spring size, lengthen pushrods, and so on. Head diameters will be 1.770-inch on the intakes and 1.500-inch on the exhausts, and they will use the late-type keeper grooves, which are the same as the Chevy smallblock. These valves will virtually drop into the production high-port heads, and would work very well with moderate lift cams, and with wilder

grinds if you use roller-tip rockerarms.

Shaft-mounted rockerarms are usually a straightforward, rugged design. Not so in the Buick V-6. The early, cast aluminum rockers are fairly tough and they are of 1.6:1 ratio (later rockers are 1.55:1), but they are non-adjustable and they take 5/16-inch pushrods. Using large-diameter pushrods in these rockerarms will chew up the sockets. Furthermore, these rockers came on engines which oiled through the rockershafts; if used on '75 and later engines which oil through the pushrods, the rockers will not receive sufficient oiling, especially at idle.

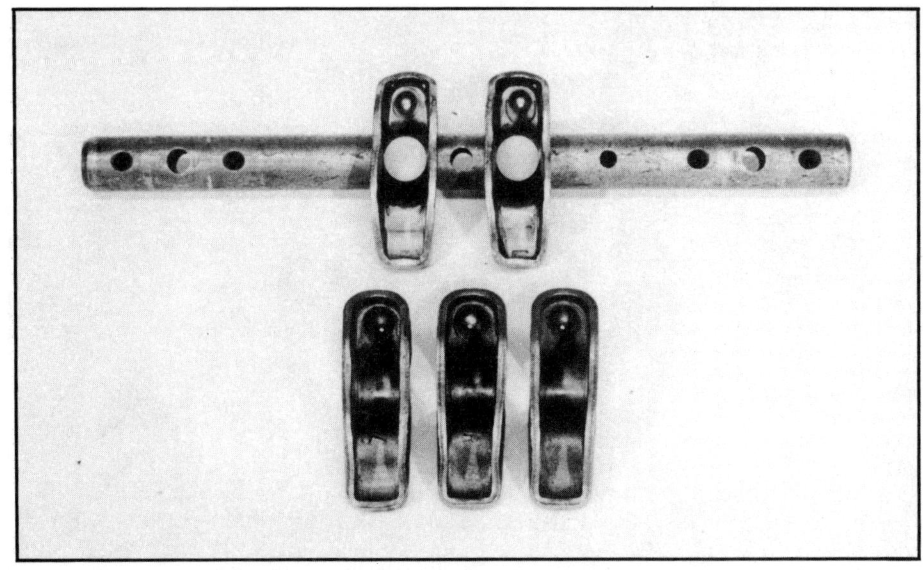

The stock rockerarm setup is definitely a liability when it comes to building for performance; and the '75-up design, with stamped rockers, held in place by push-in plastic buttons, is the worst. The '75-76 heads have bolt-on towers to support the rockershafts, and are fitted with "left" and "right" rockerarms with offset pushrod pockets. The '77-up heads have integral shaft towers, and use identical rockers, with centered pushrod pockets, for intake and exhaust.

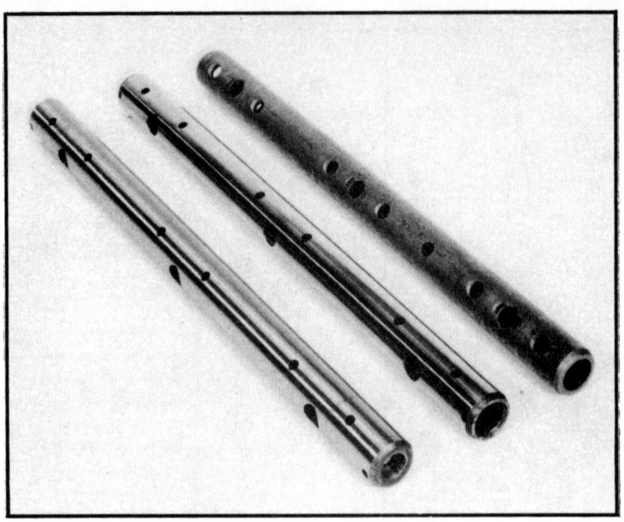

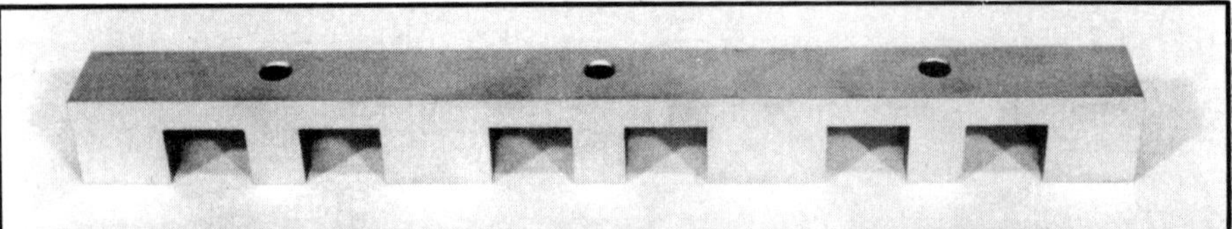

(Replacement aluminum rocker and shaft sets for pre-'75 engines are available from TRW as part 40003, as are compatible pushrods, part 48002.)

The stamped steel rockerarms and teflon retaining buttons used on '75 and later engines are good enough for use below 6000rpm, with less than .490-inch lift cams and 100-pound (closed) valve spring pressure. Beyond that, you're going to need something stronger.

The second problem is the rocker-shaft itself. Production shafts are made from relatively thin tubing, and they are very prone to cracking at the center bolt hole if used with high-lift cams and high-pressure valve springs. In the past some builders have cured the problem by making their own shafts from 13/16-inch drill rod, but now Kenne-Bell offers heat-treated, 4130 chrome-moly, steel rockershafts (KB9401) with .236-inch wall thickness (stock shafts are .136-inch wall), plus similar thick-wall shafts have recently been offered by Baker Engineering.

These shafts should solve the breakage problem with all but the most radical drag-race or circle-track roller cams. In such cases, the culprit is the shaft mounting design, with only three pedestals, which leaves the end rockers unsupported and causes flexing of the shaft. Kenne-Bell has recently designed a unique "over-the-top" girdle to help alleviate this flexing by holding the ends of the shaft stationary. Jim Bell notes that this girdle is needed only on 8000rpm-plus motors, or those using very radical roller or mushroom cams for drag or circle racing.

Smokey Yunick, among others, experimented with converting the Buick heads to accept Chevy-type stud-mounted rockerarms, but concluded that the considerable welding and machining involved (cutting off the rockershaft pedestals opens the water jacket) ultimately wasn't worth the effort. Others, like Jim Ruggles, have adapted several varieties of rockerarms to the shafts with varying degrees of success. Since the small diameter of the Buick's cam journals limits lobe lift, and you can get into oil leakage problems at the top and bottom of lifter bores with very high lifts, he likes to restrain cam profile and use increased-ratio rockerarms instead.

Ruggles has used 1.7:1 Chrysler smallblock and 1.76:1 Ford 427 shaft-type rockers by bushing them to fit the Buick shafts. To make them work on the Buick heads he mills the rocker pedestals flat at the top, then opens the shaft-mounting bolt holes from 5/16- to 3/8-inch, re-indexing them at the same time to move the shaft about .125-inch towards the pushrod side of the head. He then mounts the shafts in pillow blocks made from cut down Olds F-85 aluminum V-8 rockershaft pedestals (which have larger bolt holes than the Buicks).

Moving the rockershaft not only realigns rockerarm geometry, but also gives more room for large-diameter valve springs and retainers. At the same time, he can machine the shaft-mounting blocks to the correct height to compensate for long-stem valves. Using .100-inch longer Chevy valves, Ruggles increases shaft height by .220-inch. With such rockers, provision would also have to be made to oil them through the pushrods. Ruggles installs adjustable pushrod pockets in them from Crane or Kenne-Bell al-

Aluminum roller-tip rockers solve many valvetrain problems in the Buick, including allowance for valve adjustment and pushrod or valvestem length variation. Shown here mounted on an early stock Buick shaft are the lightweight and relatively inexpensive Crane rockers.

uminum rockerarms. Another possibility would be to convert the top end to oil through the rockershafts.

Although such modifications to the rocker system have their advantages, the task of strengthening and improving the Buick top end for high-performance has finally been greatly simplified by the recent introduction of aluminum roller-tip rockers specifically for the Buick from Crane Cams and from Kenne-Bell. The Crane rockerarms feature 1.6:1 ratio and come in a set with rockershafts of slightly higher strength, but the same wall thickness as the stock Buick's. The Kenne-Bell rockerarms (KB9301) are 1.6:1 ratio, are somewhat beefier in design, and are made to fit the tough K-B shafts (they will also fit on pre-'75 stock shafts, but not on '75 and later). The K-B rockers will work on either through-the-pushrod or through-the-shaft oiling systems.

Of the two rocker designs, the Kenne-Bells appear to have an advantage in strength and reliability. In either case, stronger rockershafts should be used, and they should be attached with high-strength bolts. Kenne-Bell also offers larger 3/8-inch diameter studs to retain the rockershafts (KB9507—complete shaft installation kit).

CAMSHAFT AND VALVETRAIN

EVEN- AND ODD-FIRE CAMS

For engine builders (and camshaft manufacturers) the Buick camshaft and valvetrain design poses several peculiarities. To begin with, all of the accessory drives on the cam are located at the front (distributor/oil pump gear and fuel pump eccentric). Through mid-1977 the camshaft drive gear, fuel pump eccentric, and distributor drive gear slipped onto the nose of the cam, in that order, and were held in place by a keyway on the shaft and a locking bolt and washer on the nose. Mid-'77 and later cams have an integral fuel pump eccentric and distributor gear, with a large flange to which the cam-drive sprocket attaches with two bolts. Both types of cam were also fitted with a more or less ineffective small spring and plastic button "thrust bumper" which rides against the inside of the timing cover. Incidentally, the '77 and earlier "take apart" cam requires a different "bent-arm" fuel pump than the later cam. Some builders, such as Ray Baker, prefer the take-apart cam design, and most cam grinders can supply either even- or odd-fire cams in this style.

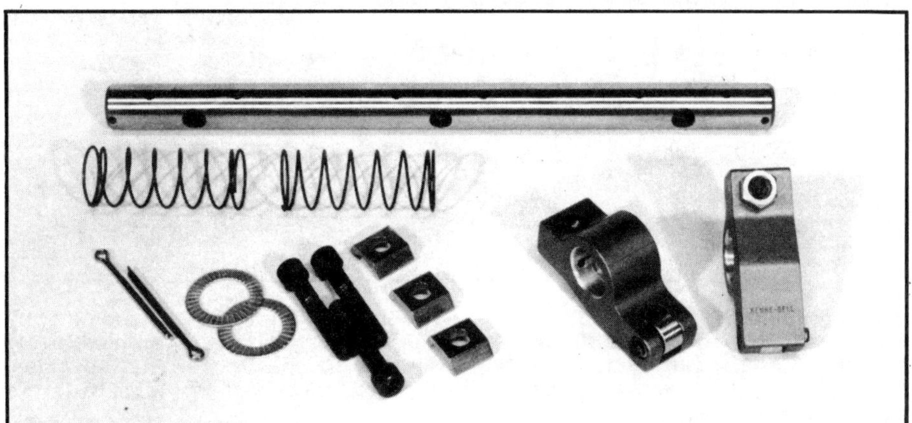

Kenne-Bell roller rockers are a bit beefier and are designed to oil either through the pushrods or through the shafts. Complete kit also includes heavy-duty rockershafts.

Another major difference in Buick V-6 cams is the change in lobe spacing between odd- and even-fire motors. All production cams with the take-apart front gear assembly are for odd-fire motors. Although Buick switched to the even-fire configuration during 1977, the cam change came a few months earlier. The majority of in-

On Mid-'77 and earlier Buick cams the timing gear, fuel pump eccentric and distributor drive gear slip over a keyway and are held in place by a bolt and retaining washer. Mid-'77 and later cams have an integral eccentric and distributor gear, and the timing gear bolts onto a large flange. Both types come with ineffectual bumper buttons on light springs.

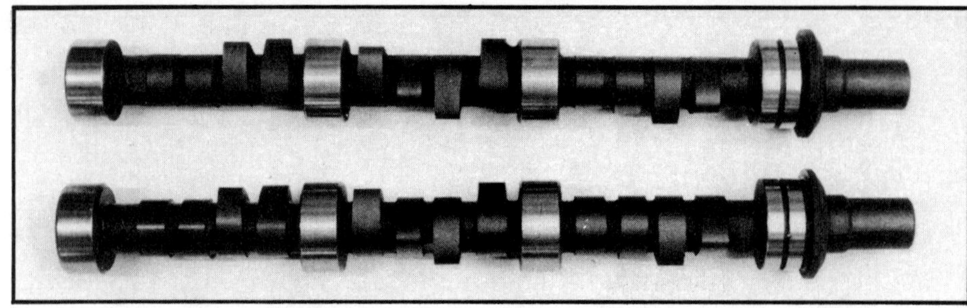

It is very difficult to tell the difference between even- and odd-fire cams just by looking at them. It is also difficult to spot a pre-'75 cam with staggered diameter bearing journals, but it makes a big difference to your motor. Of these two aftermarket cams, one is an odd-fire with stepped journals, the other is even-fire with equal journals.

tegral-gear cams are even-fire, but some were made in the odd-fire pattern; these have casting number 1254203 on the shank. Aftermarket cams have been made in both designs in both even- and odd-fire patterns, however. If such a cam is not marked even- or odd-fire, one way to tell the difference is to measure the splay angle between the front two lobes. On odd-fire cams this angle should be 39 degrees, on even-fire it is 24 degrees (given some variation for changes in lobe centers on aftermarket cams).

A third difference in production cam design was a change in 1975 to all equal size cam journal diameters (1.785 inches). Pre-'75 Buick and Jeep V-6's have stepped cam journals and bearings, decreasing in size from front to back. 1975 and later cam bearings cannot be installed in earlier blocks, and later cams will not fit in earlier engines unless the journal sizes are cut down.

These latter two changes, though not readily apparent to the eye, make a big difference in fitting the correct cam to a specific V-6. There have been cases of odd-fire cams installed in even-fire motors, with predictably dismal results.

PERFORMANCE CAMS

Aftermarket high-performance and RV-type cams are readily available from most cam companies for the Buick. However, far too many cam grinders live almost exclusively in a Chevy smallblock world, adapting their Chevy technology to the "off brands." One exception is Crane Cams. They have worked closely with Smokey Yunick and other builders on specific Buick V-6 projects, and they offer several good profiles for the Buick, as well as several special Buick V-6 valvetrain pieces. As one of the ingredients of his "5-step" plan, Smokey recommends the Crane H-214/2867-12 hydraulic grind which features 214 degree duration (at .050-inch lift), 112-degree lobe centers, 65 degree of overlap, and .464-inch intake and .494-inch exhaust lift with 1.6:1 rockerarms. This cam has a strong 2500 to 5500rpm power range with approximately 7000rpm maximum capability. This cam is designed for even-fire motors, and would work well on street applications. For his early testing on the odd-fire motor (which pulled 350-plus horsepower), Smokey used cams in the 270-280 degree range with lifts close to .600-inch, and even tried different profiles and lobe centers for the cylinders firing at 150 degrees from those firing at 90 degrees.

In the realm of Buick cam technology, however, one company has a decided advantage in terms of experience and specialization. Kenne-Bell has been grinding cams for Buick V-6's for nearly 15 years, and they make cams only for Buicks. Their current catalog lists several profiles for the V-6, ranging from very effective NHRA-legal "cheaters" to radical rollers, with a full complement of valvetrain hardware. Their more popular grinds include the KB Mark 2, which they describe as an excellent "street, strip, off-road, grocery-getter" cam which pulls to 6000rpm and still de-

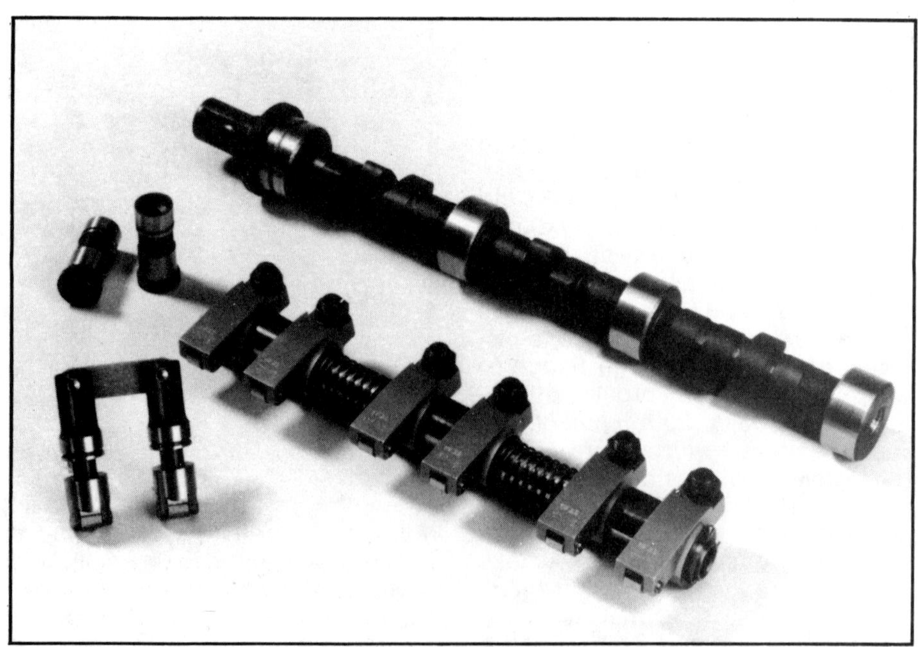

To achieve valve adjustment without changing rockerarms, you could resort to adjustable pushrods (right), but it's far from a perfect solution. For any high-performance situation, it would be advisable to switch from stock 5/16-inch to 3/8-inch pushrods (middle). Pieces shown are from Kenne-Bell.

Besides Kenne-Bell, Crane Cams is one of the few companies that offers a full line of high-performance valvetrain components for the Buick V-6, including roller rockers, their unique oil-control roller lifters, and mushroom solid lifters.

Lineup of valve springs from Kenne-Bell includes (left to right): stock spring, KB11001 performance spring and damper which fits the V-6 without modification, Buick-size dual spring, KB10702 high-performance dual spring which fits V-6 heads after trimming valve guides, and super high-pressure triple spring for roller cams, which may require notching undersides of roller rockers for clearance.

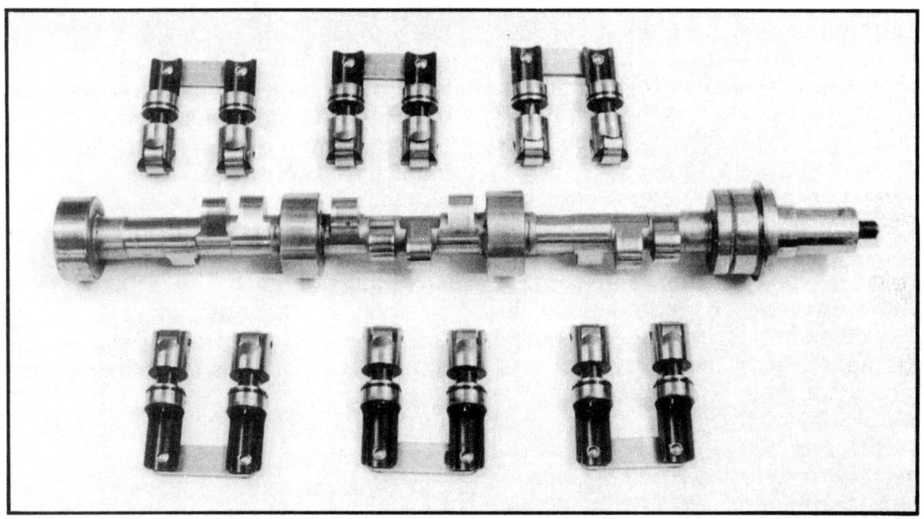

Typical billet roller cam is shown with Crane lifters for V-6's. Note how cam is cut to small base-circle diameter to get high lift, while nose of lobes nearly equals bearing journal diameter.

livers slightly better than stock mileage. This is a dual-pattern cam, designed to help scavenge the exhaust ports on pre-'79 heads. The KB Mark 3 is also a dual-pattern with a power band from 3000 to 7500rpm with an idle speed of 1100rpm. It requires notched pistons and is not recommended for street use in automatic-shift cars.

LIFTERS

Jim Bell strongly recommends hydraulic lifters for the majority of Buick V-6 applications. Since all Buicks came with hydraulics, and have no provision for rockerarm adjustment, it is most practical to stick with hydraulic cams for street or moderate performance build-ups. Kenne-Bell makes a Hi-Rev hydraulic lifter expressly for the Buick (KB10001 for '61-69 engines, KB10002 for '75 and later) claimed to be good to 7500rpm, as well

as Buick-length lightweight chrome-moly, 3/8-inch diameter pushrods (KB10201).

Since Chevrolet lifters have the same bore diameter as Buicks, most aftermarket cam companies recommend "universal" (i.e., Chevy) performance-type hydraulic lifters with Buick V-6 cams. The only problem is that the pushrod seat in Chevy lifters is .100-inch lower than in Buick lifters, which means that Buick-length pushrods will no longer work. If the base circle of the aftermarket camshaft has been ground smaller to increase lobe lift, this will only compound the problem.

Several of the established cam manufacturers (such as Crane) market "engineered kits" for the Buick using Chevy-type lifters plus high-performance pushrods made to the corrected length. Since Buick V-6's come with 5/16-inch diameter pushrods, it would be smart to install stronger 3/8-

inch pushrods with a higher performance cam anyway, so buying a matched cam and kit would be fine. But if you start putting together your own mix-n'-match valvetrain assembly, be advised that Chevy parts aren't necessarily the same as Buicks. Jim Bell also cautions that lifter-to-bore clearance is critical in the Buick V-6. Whereas some manufacturers like to cut lifters for looser tolerance, Bell states that decreasing lifter diameter by just .001-inch can bleed enough oil from the main gallery (remember it runs past all the right-hand lifters) to drop engine oil pressure 10 to 20psi.

TRW lists a good selection of performance hydraulic cam and valvetrain components for the Buick which you should be able to order through your local parts store. For R-V-type street performance they recommend camshaft TP-188 which features a high-lift, short-duration profile (.424-inch, 270° intake; .448-inch, 280° exhaust). TRW VS-966 (.495-inch maximum lift) or VS-1008 (.520-inch lift) performance valve springs would work well with such a cam and can be used with the stock retainers. They should be installed at 1.720-inch height. This cam is compatible with stock lifters (TRW part VL73) and rockerarms, though aftermarket, heavy duty pushrods would be a wise addition.

For higher levels of performance TRW offers hydraulic cam TP-198 which has 300 degrees duration and .496-inch lift and a power band from 2000 to 6000rpm. It can be used with stock lifters and the VS-1008 springs metioned above. For use with the TRW stainless race valves and more radical (.600-inch lift) cams, TRW offers VS-1100 outer- and VS-1101 inner springs and SR-363 chrome-moly or SR-336 aluminum retainers. To install stock-diameter springs with the TRW valves, you can use SR-364 chrome-moly retainers. For high perormance use, TRW recommends aftermarket, aluminum roller rockers and heavy-duty pushrods.

Nobody makes solid lifters specifically for Buick V-6's, but pushrod length isn't as critical since some form of lash adjustment must be incorporated if you convert to a solid-lifter cam. The inexpensive method is to install a set of adjustable pushrods, available from several sources including Kenne-Bell (KB10301). However, such pushrods are not only a bear to adjust, but a system retaining the stock stamped steel rockerarms is limited to under 6000rpm operation—and you might as well use a hydraulic cam if that's the range you're shooting for. For high-rpm performance (with hydraulics or solids) your best bet is to install a set of the new Crane or Kenne-Bell adjustable aluminum roller rockerarms.

Whereas most builders have used restrictor-type pushrods (with a smaller hole in the tip) to limit oiling to the rockers in Buicks, Jim Ruggles found that using Chevy edge-orifice, solid lifters in the Buick does not meter enough oil to the top end because the oil hole does not come close enough to the gallery in the Buick. To increase oiling, he grinds a .003-inch flat on the side of the lifter from the middle band up to the oil hole. Using inertia-valve lifters instead might serve the same purpose.

VALVE SPRINGS

In the valve spring department the aftermarket is once again skewed toward Chevy components. However, smallblock Chevy valve springs are both wider (1.550-inch diameter) and taller (an extra .060- to .070-inch) than the Buicks. In the past, builders using Chevy valves in the Buick heads would also use Chevy-type springs (by enlarging the spring seats in the heads) because of their availability and to maintain the correct installed spring height with the longer valves. However, use of these larger diameter springs and retainers sometimes requires grinding the undersides of aluminum rockerarms (stock or aftermarket) for clearance.

To simplify installation with stock-length valves (either Buick or the new performance valves scheduled from TRW), Kenne-Bell offers a dual valve spring set (KB10701) in the standard Buick diameter and installed height.

They require trimming of the pad with a special cutter (KB11401) to position the inner spring, but this is a simple operation accomplished with a hand drill. They also require special KB-11102 steel retainers, which are cut to a close .003-inch tolerance fit in the springs to eliminate any lateral movement of the spring on the retainer. These springs provide 140 pounds installed seat pressure and are compatible with cams of up to .600-inch lift. As mentioned earlier, '75-up heads will also require trimming of the tops of valve guides with cams over .417-inch lift when using stock-length valves.

ROLLER CAMS

We have already discussed some of the peculiarities of installing roller lifters in the Buick (see oiling section). Crane currently markets the only rollers made specifically for this motor. They have an extended nose partially covering the roller on one side to keep oil from escaping from the enlarged lifter (main) oil gallery. These lifters are designed to work in the stock lifter bores. However, due to irregularities typical in block castings (core shift), as well as some degree of variation in the size of the nose (due to machining tolerances) from lifter to lifter, it is imperative that complete oil passage sealing *on each lifter* be physically checked with the cam and lifters installed in the block. The potential for oil leaks at the top or bottom of these lifters increases with high lobe lifts, especially if the cam must be ground to a smaller base circle to attain them.

The problem is, how do you check them? A method used by Jim Ruggles is to remove the small press-in plugs at the front of the block sealing each lifter

gallery. Then, with the cam and lifters installed, he reaches into the gallery with a piece of long, stiff wire with a short right-angle bend at the end. With this probe he can "feel" each lifter at the top and bottom edges of the gallery opening, as the cam is turned through its cycle by hand, to determine if the head or nose of the lifter will drop or lift far enough to open the oil passage.

A further method is to spray each lifter with machinists' bluing and use the wire to scribe a line at the lifter's highest and lowest lift points. If you discover lifters which will allow oil to escape, you might be able to swap a short lifter into a longer bore, you may have to replace the lifter, or you might even have to find a better block. Also remember that lifters which are very close to the limit at maximum lift are likely to open the passage if there is any separation of the lifter from the lobe at high rpm (a situation which you ultimately want to avoid, anyway).

Another alternative is to sleeve the lifter bores. However, doing so restricts the main oil gallery and an "outside" oiling system must therefore be incorporated to feed the main bearings. Kenne-Bell uses this system on their K/Gas drag motors, using .050-inch wall bronze sleeves drilled with .030-inch holes for oiling to the lifers. The sleeves and a complete roller cam and kit are available from

A positive thrust bumper is a necessity for a performance cam in a Buick. Take-apart cams, which use a bolt to hold the gears on, can easily be fitted with an adjustable, positive-stop device, as shown on the right. Kenne-Bell integral gear cams are specially made to accept a simple teflon press-in button, as shown on left. In the center is the stock bumper.

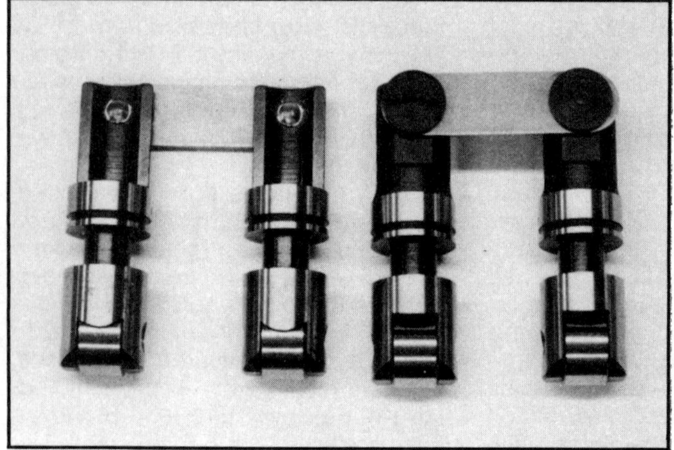

Front and back view of Crane rollers shows extended flank on one side to cover large lifter (main) oil gallery in the Buick, plus the deep groove in the lifter body to allow oil to flow through the gallery. The same lifters work in Buick and Chevy 90-degree V-6's.

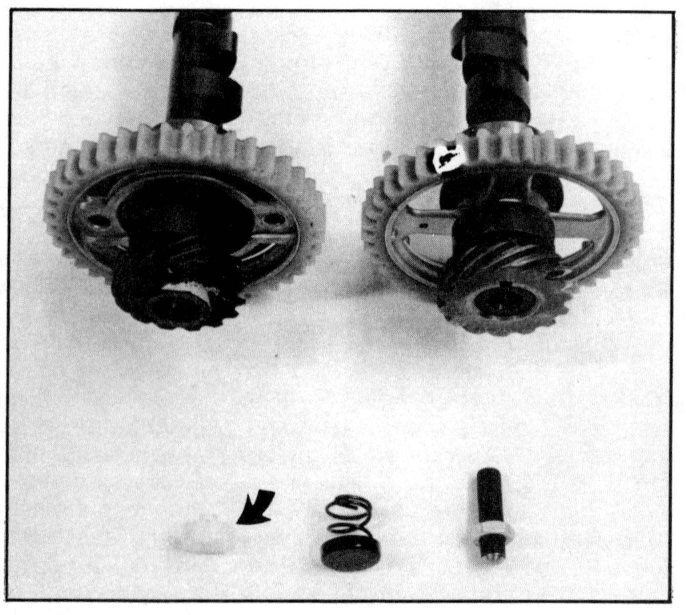

Kenne-Bell. The only problem they have encountered with this system is some breakage of the lifter bosses in present blocks. This problem is due to extreme spring pressures and the slight weakening of the lifter bosses, created when material is removed from the lifter bore to allow the sleeves to be installed. On early blocks, they resorted to filling the entire lifter chamber (valley) with Devcon (epoxy) to support the lifter bosses. This problem should be alleviated in the newest heavy-duty blocks, which have added strengthening ribs between the lifter bores.

Another disadvantage with current Buick designs for roller-cam installation is that the cylinder head has no "lip" over the valley chamber to allow the installation of a rev kit plate. Butch Elkins sees this as the major drawback to high-rpm capabilities in the Buick, and he is lobbying to have the face extended on potential Stage 2 heavy-duty heads to support a rev kit plate. As of now, however, there is no practical way to install a rev kit on a Buick V-6.

We have also mentioned the limitation to cam lobe height caused by the small cam bearing diameter in the Buick block. To get high lifts, the lobes must be cut on smaller base circles and the entire cross-section of the cam will be reduced, which leads to possible deflection and torsional twist, especially with excessive spring pressure. The new heavy-duty blocks have increased material around the cam bearing bulkheads, so they could be bored out to accept smallblock Chevy cam bearings. However, Jim Ruggles notes that enlarging the cam bearing size in the Buick block will intersect deeper into the current oil passages to the main bearings, restricting flow, so such a modification would have to be accompanied by an external oil feed to the mains.

Much of the above discussion is directed to the most serious racers, who are a small percentage of the readers of this book. Valvetrain specifics for all out competition versions of the Buick V-6 will vary widely, especially in this period of experimentation while builders are trying all sorts of hybrid components to find the best combinations for making lots of reliable horsepower with this motor. Although it's been around for years, the Buick V-6 is still going through an adolescence as a high-performance powerplant, especially with all the new heavy-duty parts coming from the factory. If you are building a radical Buick, the major considerations in

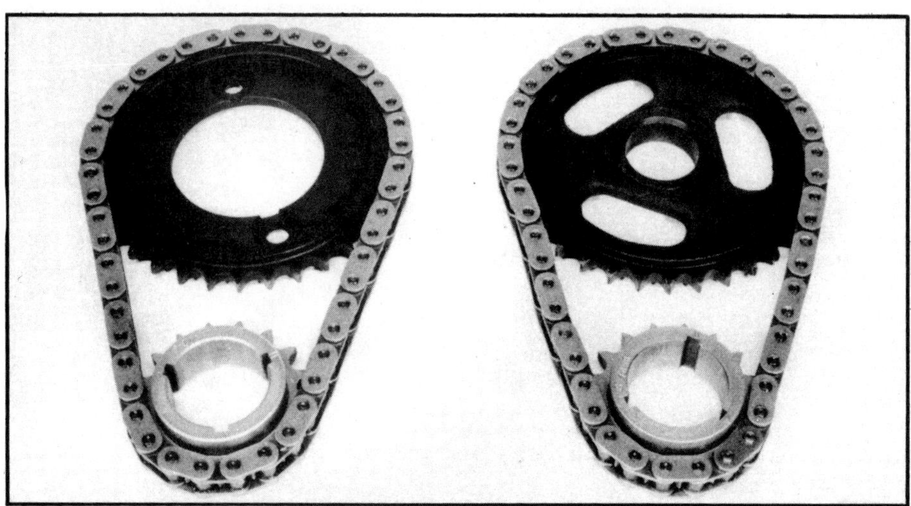

Cloyes single-roller timing gear sets are available for both styles of camshafts, and are a considerable improvement over the stock chain and nylon gear. Note that the crank gear is cut for advance or retard cam settings, as well.

There's nothing terrible about the pre-'78 stock 2-barrel intake. It has been used very successfully in those IMSA and NHRA classes that are limited to 2-barrels. These manifolds and the Rochester carbs that go with them can't be beat for the price—dirt cheap.

valvetrain components are stronger rockershafts and rockerarms, correct rockerarm geometry, proper pushrod length, clearance between valve springs and rockerarms, and sufficient (but not overabundant) oiling to the rockerarms on the shafts. New compatible components such as stock-length performance valves, springs, and lifters plus adjustable roller rockers and strong shafts now make valvetrain upgrading much easier on the Buick V-6, even for backyard engine builders.

If you install a performance cam in the Buick, you should also install a positive-stop thrust bumper in place of the stock spring. This inexpensive piece will limit cam walk, reduce spark scatter in the ignition, and improve timing gear life by maintaining correct alignment. Kenne-Bell offers an adjustable-type (KB9901) for '77 and earlier take-apart cams (adjust for .004- to .008-inch endplay). Their own integral-gear cams come with a simple teflon button which presses into the front of the cam, but they only fit K-B

cams, which are machined to accept them. But it wouldn't be difficult to fabricate your own inexpensive positive-stop bumper.

Timing chains also seem to be a weak part in Buick V-6's, especially on odd-fire engines which impart uneven harmonics on the chain from the crank gear. The problem does not lie in the stock nylon cam gear, but rather in the stock-type chain which stretches and "goes away" with regularity. Although not as efficient as their double "true-roller" chains for other engines, Cloyes makes single-roller timing chains and gear sets for the Buicks, which are available from Kenne-Bell. Complete set KB17401 (Cloyes 9-3128) is for take-apart cams, KB17402 (Cloyes 9-3129) is for '77-up integral-gear cams. These timing sets aren't optimal, but they are better than the stock system, and they are the only alternative currently on the market. If you are using a stock link chain, be sure to inspect it frequently and replace it when necessary. Some builders actually prefer the nylon cam gear,

The Rochester 2GC carb has been used in hundreds of applications over the years and comes in a variety of sizes. The '75 Buick used a 298cfm version, while '76-78 models had 205cfm ratings, though both had the same 1.697-inch throttle bores (the '75 carb has larger venturis). However, even bigger versions of the 2GC exist, as demonstrated by the one on the left, and can be bolted onto the stock intake.

saying it helps absorb crankshaft vibrations. Avoid steel replacement cam gears for use with stock chains; a large batch which were not properly hardened seem to be lingering in the market.

INTAKE MANIFOLDS AND CARBURETORS

STOCK INDUCTION SYSTEMS

In 1978 Smokey Yunick termed the stock Buick 2-barrel intake "below horrible" and proceeded to fabricate his own open-plenum, hi-rise manifold for it. A copy of this manifold is now being marketed in cast form by Weiand, with Smokey's name on it. According to Smokey's dyno, this intake coupled with a Holley R-4776-2 (600-cfm) 4-barrel will yield an 18 horsepower increase over the stock Quadrajet 4-barrel induction system on a production 4.1 engine. This is the first increment of Smokey's "5-step" plan for the Buick. In actual testing of

Smokey's recommended modifications conducted by *Hot Rod Magazine* on a 4.1-liter engine swapped into a 1980 Skylark (X-body), the Weiand manifold and Holley carb dropped the quarter-mile E.T. from 14.97 to 14.64 and increased mph from 91.27 to 93.96, over the stock 4-barrel manifold and carb. For this test, however, the tops of the manifold runners were ground out to match the high-port heads.

By comparison, Kenne-Bell was campaigning an NHRA legal R/Stock 1975 Buick Skyhawk several years ago which ran 14.20's at 96 mph with a 3.8-liter, odd-fire motor using the *stock* cast-iron 2-barrel intake manifold and a re-jetted Rochester 2GC 2-barrel carb. This 2800-pound machine had the advantage of 4.10 rear gears and 9-inch slicks, plus slightly higher compression (about one point), but it also had 21 fewer cubic inches than the Hot Rod Skylark. The point is that the early manifold and Rochester 2GC 2-barrel (used on 1961 through 1978 Buicks)

aren't "horrible." Admittedly, Smokey was referring to the later "spread-bore" 2-barrel intake, which uses a Rochester E2ME carburetor similar to the primary side of a Quadrajet with 1.375-inch throttle bores and 1.218-inch venturis. This induction system was put into production on 1979 models along with the switch to high-port heads, and the manifold flows rather badly.

Unfortunately the pre-'79 2-barrel intake with the familiar and inexpensive Rochester 2GC carb won't fit the newer heads. All of the 2GC carbs used the same throttle bore size (1.697-inch), but the 1975 version had considerably larger venturis than other model years (1.250-inch, rather than 1.093-inch). This carb is rated at 298cfm (as opposed to 205cfm for '76-78 versions) and the 1975 engine is consequently rated at 110 horsepower, dropping to 105 horsepower the next year. Bolting this carb (part 7045145) onto other pre-'79 motors might get you a quick and cheap five horsepower. Another model of the 2GC, which came on mid-'60's Chevy pickups, is rated at 330cfm and will bolt onto the early 2-barrel manifold. These carbs can be ordered from Kenne-Bell (KB47000).

The stock, aluminum 4-barrel intake, which was introduced as standard on 1980 4.1-liter cars (Rivieras, Electras, and some LeSabres) shouldn't be shortchanged, either. This setup was used on the 1981 Indy pace car Buick to produce 280-300 horsepower with a relatively mild hydraulic cam. According to Jim Ruggles, who assisted on the program, "The stock 4-barrel is a damn good manifold." By simply swapping this manifold and carb for the production 2-barrel on an otherwise-stock 3.8-liter motor (rated at 105hp at 3800rpm), Ruggles saw 138 horsepower at 3200rpm on his dyno. The stock, 4-barrel intake features a low profile (which gives needed hood clearance on many late-models), a full center divider, and it matches the late high-port heads. It might not be any less expensive than an aftermarket manifold and carb if purchased over the counter from Buick, but the price could be very tempting from a wrecking yard. Besides better hood clearance, Ruggles points out that the stock 4-barrel is excellent for street applications because it has the center divider and a small balance tube which give better low-end response than most aftermarket intakes with large open plenums. The drawback is that it will only fit the '79 and later heads.

For installing this or other "heated"

manifolds on the high-port heads, Ruggles suggests using the 1980 Turbo manifold gasket (Buick part 25504064) which has a smaller 5/8-inch hole covering the large exhaust heat crossover passages. This small change significantly reduces pinging in the engine without totally eliminating carburetor warming for practical street driving.

Buick greatly increased the size of the exhaust heat passages in the heads in 1975 to promote faster carb heating for emissions control. On a race engine these passages should be blocked off completely. Kenne-Bell sells plug kits for this purpose for the early heads, plus they have just introduced a new intake manifold gasket for the high-port heads which completely blocks the heat passages (KB44300).

Street engines should have some manifold heating for practical carburetor operation. You could use the KB gasket and drill 5/8- or .50-inch holes in it for adequate heating, or use the 1980 Buick turbo gasket. Jim Bell also notes that, although some companies have introduced composition-type manifold gaskets for the V-6, the stock steel shim gaskets are preferable since they are very durable and incorporate a valley shield to keep hot oil off the bottom of the intake manifold.

AFTERMARKET INTAKES

Aftermarket intake manifolds are plentiful for the Buick V-6. Offenhauser, as usual, got into the act immediately, designing a "360° Equa-Flow" intake for the first 198-inch Buick back in 1962, in styles to accept either Holley/AFB or Quadrajet 4-barrels. They still offer this manifold, plus their patented "Dual Port," in several 4-barrel configurations to fit pre-'79 (small-port) engines. Offy has also recently brought out a new "X-type," open-plenum, 4-barrel intake specifically for the Buick, in different versions to fit each of the V-6's from 198's through the current high-ports. This manifold features a raised floor which acts as a partial divider, with a water passage underneath for manifold heating (it can be blocked off by a screw-in plug).

Weiand, Holley, and Kenne-Bell each offer more or less similar single-plane, 4-barrel manifolds for the Buick. The Holley "Street Dominator" (part 300-22D) is currently available only for '75-78 engines (low-port heads). It has a full center divider and is designed for use with the small 450cfm Holley model 4360 carburetor, though it will

The new factory aluminum, 4-barrel intake manifold, mated with a Quadrajet, comes on 4.1-liter motors and will fit any '79-up heads. It's an excellent performance swap for the 2-barrel, especially if you can get it from the wrecking yard.

accept other Holleys, AFB's, and Quadrajets. Of the three, this is the most street-oriented manifold, typical of the Street Dominator line, aimed at performance and mileage improvement from idle to 5500rpm.

Weiand offers two versions of the Buick intake, both developed by Smokey Yunick. Model 7541 is a street version with standard heat, water, and EGR passages, plus a removable carburetor pad to accept various carbs. They recommend fitting it with a 600cfm Holley (model 4776 or 8004). This manifold is a bit taller than the Holley and has a fully open plenum. The Weiand model 7542 is an all-out race manifold designed for competition in the 7000rpm-plus range. It has an open plenum, raised runners with an air gap underneath, a separate water housing at the front (which might cause some interference with HEI distributors), and no heat passages or other street hook-ups. This manifold is obviously impractical for the street; it would be excellent for circle-track competition.

The new Kenne-Bell manifold (KB-44100) professes to be a combination street and race intake giving increased torque and economy at lower speeds plus power potential through 9000rpm. It has an open plenum, reduced heat passages, and provisions for standard street attachments. Both the Kenne-Bell and the Weiand manifolds feature an "intermediate" port size which will fit either low- or high-port Buick heads. As Jim Bell explains, "The heads are the problem for flow, not the manifold. Most manifolds flow more than the heads." Thus a smaller manifold runner and port can be mated to the larger head port, if the angle of the manifold runner is designed to direct the flow efficiently into the port. For street applications, Bell

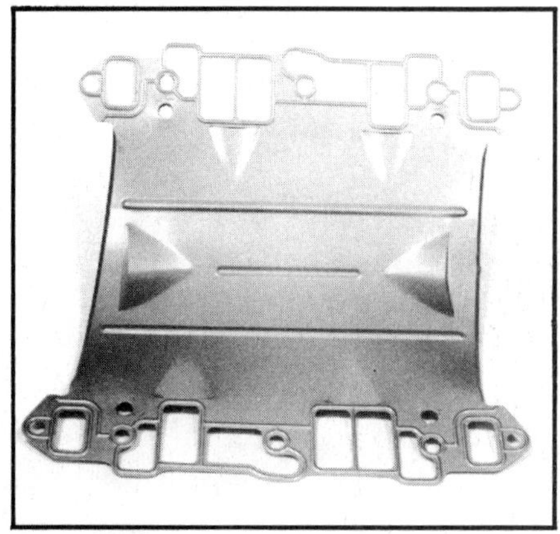
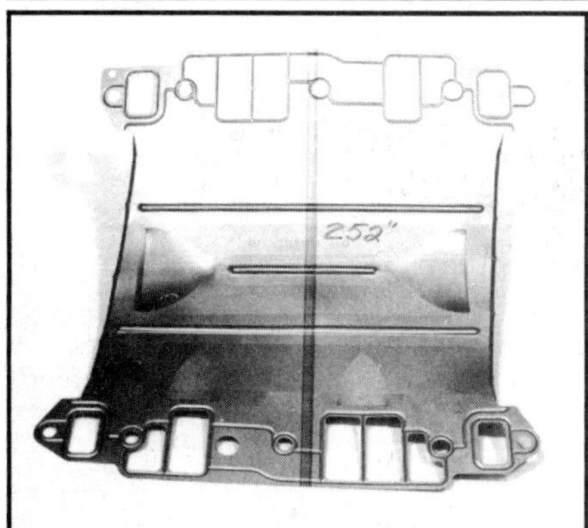
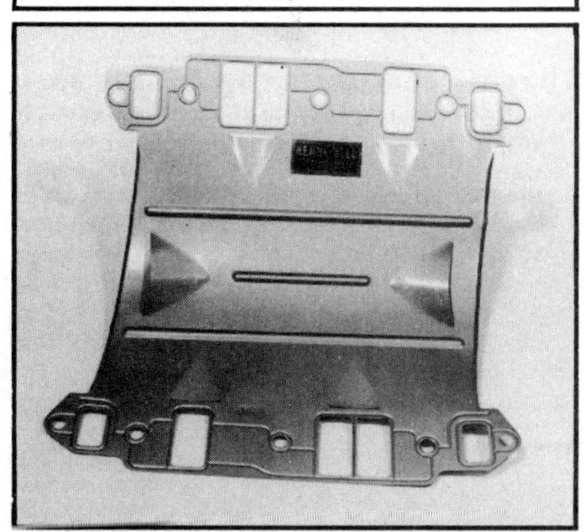

The stock pressed-steel intake manifold gasket (top) is tough and incorporates an oil-splash heat shield, but it lets too much exhaust heat to the carb. The 1980 turbo-motor gasket (center) has a smaller heat passage hole, giving street motors some free horsepower. Kenne-Bell now offers a heat-blocked gasket (bottom) for competition engines.

says not to match the manifold to the head ports. However, both the Weiand and the K-B intakes have enough material above the runners to open them up on the top side to match the high-port heads for racing. Grinding the manifold runners on the lower side would be a waste of effort in either case.

If you are looking strictly for economy gains, along with a little better performance in street driving ranges, you can install an Edelbrock S.P.2-P. (part 5186) on '79 and later high-port head engines. It is made in 2-

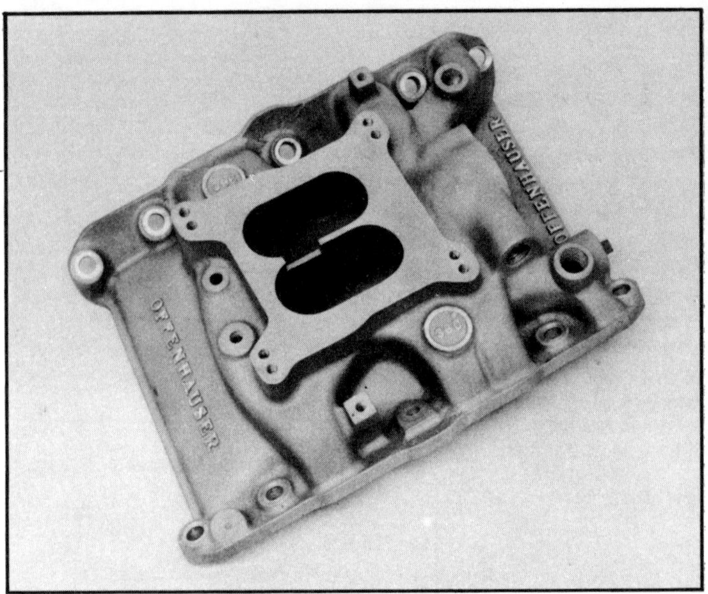
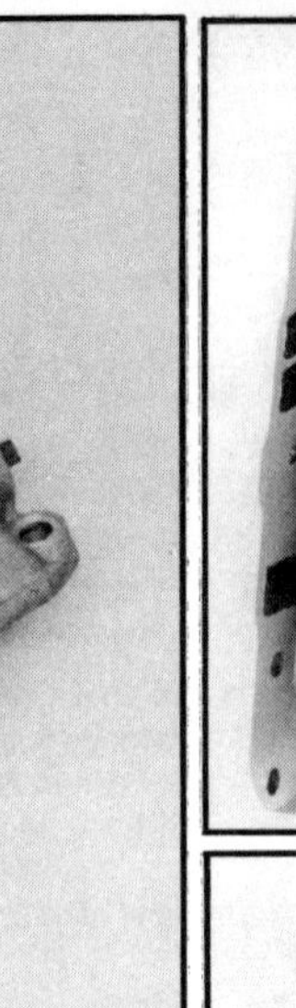

Offenhauser has been making manifolds for the Buick V-6 since 1962, and they offer a wide range of styles. Their 360-degree, 4-barrel model (top) comes in either standard or Dual-Port configurations (upper right). Their latest X-type, single-plane manifold (right) features a steep floor with optional water heating passages inside.

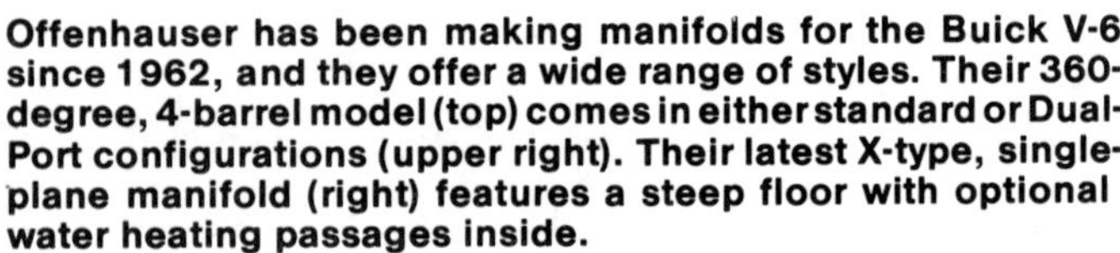
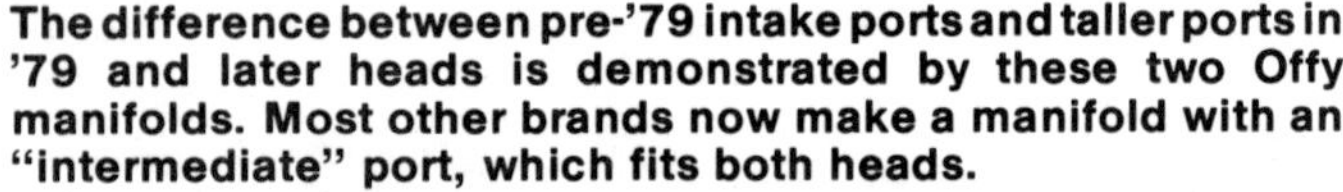

The difference between pre-'79 intake ports and taller ports in '79 and later heads is demonstrated by these two Offy manifolds. Most other brands now make a manifold with an "intermediate" port, which fits both heads.

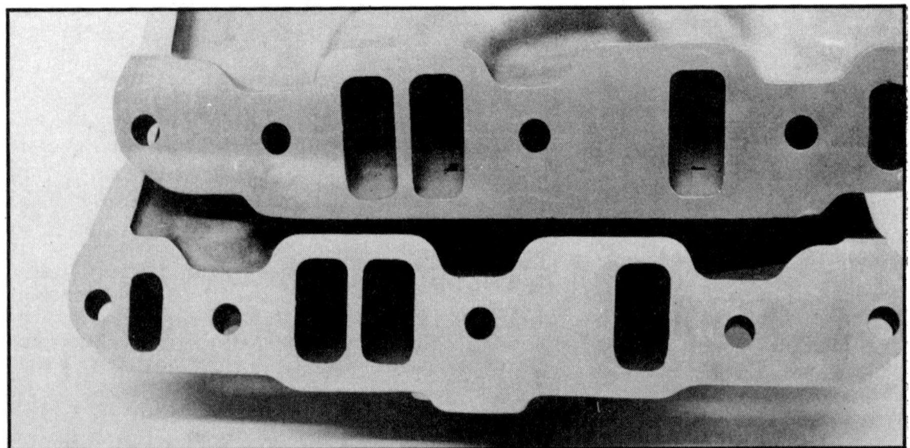

barrel version only, which accepts the stock spread-bore carb. Considering the design of the stock, late-model, 2-barrel manifold, the S.P.2-P. would be a practical addition to an otherwise-stock V-6.

As far as more exotic natural-aspiration systems for the Buick are concerned, the only equipment now available is a direct-port fuel injection system designed several years ago by Hilborn to fit the early (low-port) heads. Kenne-Bell is now using a set of these injectors, modified to fit high-port heads, on their gasser. Hilborn may incorporate these modifications in a production injector set for late-model Buicks if there is sufficient demand. Previous to switching to injection, K-B used a hand-made, tunnel-ram manifold which used three bores each on a pair of modified Holley 4-barrels. Other drag racers may find similar ingenious solutions for "ram inducting" a Buick V-6.

IGNITION

EVEN- AND ODD-FIRE FACTORY DISTRIBUTORS

All pre-1975 Buick V-6's came with a standard Delco points-triggered ignition system. Although the ignition wire terminals are evenly spaced on this small-diameter cap, the rotor has a broad terminal which maintains contact with plug wire terminals through 30-degrees of rotation, allowing sparks to be fired at irregular intervals by the odd-fire pattern of lobes cut on the breaker cam in the distributor. If a Buick V-6 is equipped with this small-body distributor, it's a good indication that it is an early (225- or 198-inch) odd-fire motor.

All 1975 and later engines came with the Delco breakerless electronic HEI ignition system, easily recognizable by the large-diameter distributor body and cap, which houses the coil in the middle. The '75-77 odd-fire motors can be readily identified by a strange-looking distributor cap. Actually a V-8 cap with two terminals blanked out, it has four plug wires grouped together on one side and two set apart on the other. All even-fire Buick V-6's came with HEI distributors which have symmetrically—though not evenly—spaced ignition wire terminals. In both cases the terminals on the inside of the cap are elongated to make contact with the rotor in the proper position for either an odd- or even-firing pattern.

98

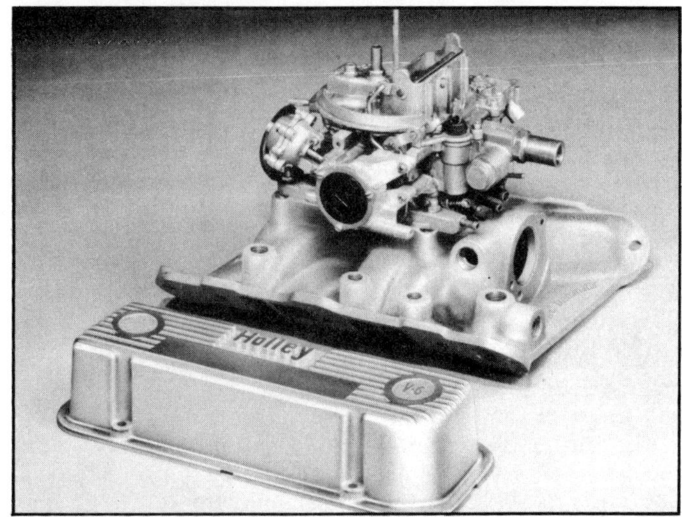

Holley's Street Dominator intake, matched with a 450cfm economy 4-barrel, is a very practical street-performance combination for the Buick.

The new Kenne-Bell Buick intake is claimed to combine excellent street as well as top performance characteristics.

The exact timing of the spark is of course controlled by the position of the heads on the magnetic pickup module in the distributor. Spotting the difference between the two HEI distributors is a quick way to tell even- from odd-fire motors.

However, all Buick V-6 distributors will interchange, and an engine will run (through not very well) with the wrong ignition. A second indicator is the front cover. All odd-fire motors have a timing mark on the front cover; even-fire motors use a bolt-on timing indicator. There is an 18-degree difference in the top dead center (TDC) mark between the two. These components are interchangeable, so it is obviously imperative to match parts for proper phasing and timing of the ignition.

If you have an odd-fire engine, you can choose between the stock points-type distributor or the HEI, plus several aftermarket systems. For street or sub-6000rpm performance, either stock system works well, with the HEI delivering a much fatter, more reliable spark in street-driving rpm ranges.

Voltage appears to drop off significantly in the HEI system above 6000rpm, but its effectiveness can be stretched another 2000rpm simply by gapping plugs at .030-inch (this applies to even- or odd-fire motors).

The odd-fire's uneven spark timing does cause problems with either points-type or capacitive-discharge ignitions (such as the HEI) in the higher rpm ranges. In a conventional ignition the intensity of the spark is partially dependent on the length of time the coil is allowed to build up voltage (or "saturate"). In the odd-fire, build-up time alternates between short (90 crank degrees) and long (150 degree) periods, which can lead to weaker ignition in half the motor at higher speeds. Electronic ignitions, which produce higher voltage in much less time, eliminate this problem. However the stock HEI is less than perfect at high rpm levels, and when "cured" with an aftermarket coil or amplifier system, the irregular firing pattern can lead to cross-firing or ionization problems in the cap. We're talking 6000-10,000-

rpm racing conditions, remember. For street or moderate performance the odd-fire shouldn't cause major ignition problems. But for racing, the even-fire configuration poses far fewer problems.

Since the majority of you, realistically, will seldom see 6000rpm-plus use with your V-6, there's little need to invest in expensive ignition equipment. (The same can be said, of course, for lots of "speed" equipment in general.) If you have a pre-'75 breaker-point distributor, Kenne-Bell markets an inexpensive advance recurve kit (KB33101) consisting of new springs and weights to give full mechanical advance by 2000rpm. Swapping to a '75-77 HEI, which are plentiful in junkyards, should show a slight improvement in street-range economy and performance.

Most ignition or speed shops can recurve either type of stock distributor. The Buick likes a total ignition advance of approximately 34 degrees, but this figure and the exact distributor curve will depend on the "knock" tendencies of your motor. In today's low-octane world, tune by ear—advance the distributor in small increments until the engine pings noticeably under acceleration, then back it off a few degrees. Water injection or gasoline blending (50% leaded regular with unleaded premium) will allow you to run more ignition lead, which will directly translate to better street mileage and performance.

AFTERMARKET IGNITIONS

If you want to invest in aftermarket ignition equipment for the street, or if you're going racing, you now have

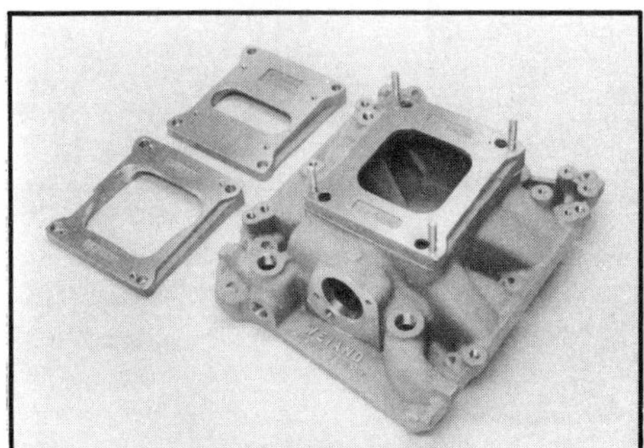

Weiand's Buick manifolds, designed by Smokey Yunick, come in a street version (left) with exhaust heat and interchangeable tops, and in a circle-track competition model, featuring elevated runners, a separate water neck and no heating.

If you are looking for maximum economy with improved performance from a '79 or later Buick, the Edelbrock S.P.2-P., which accepts the stock, late-model 2-barrel, is what you need.

Hilborn makes direct-port injectors for the early V-6 heads, which can also be slightly modified to fit high-port heads.

plenty of good choices for your Buick. Mallory makes their rugged YL-series dual-point distributor for both the odd-fire (part 2555601) and even-fire (part 2564301). Since the dual-point configuration allows greater coil saturation time, this would considerably help the odd-fire motor. The small diameter of this distributor is an asset over the big HEI if a special induction (such as a blower) is used. These distributors are also popular with racers since either is available with a mechanical tach drive (change the second digit of the part number to a 6). These units can also be used to trigger an MSD or other electronic modules, but MSD recommends a larger body (i.e., the HEI) to reduce ionization and cross-fire tendencies.

For the even-fire only, Mallory also offers a Unilite photo cell breakerless distributor (part 3764301; 3864301 with tach drive) which can be used with a standard coil or hooked up to Mallory's new multispark CD unit. Also available only for even-fire motors is Accel's BEI III breakerless distributor (part 49106; add T for tach-drive body) and Capacitive Discharge ignition system. However, Accel's new Laser I CD ignition system will supercede the BEI, and it is available in kit form to adapt to

the stock HEI distributor.

Unless you need additional clearance or mechanical tach-drive capability, there is little need to switch from the stock Delco HEI distributor for either odd- or even-fire motors. The major problem with the stock HEI is the built-in coil, and all aftermarket ignition companies now sell inexpensive kits to adapt their own high-energy coils to any HEI distributor. Kenne-Bell sells such a system, consisting of a replacement (external terminal) distributor cap dust cover, an MSD "Blaster" coil, and hook-up for odd-fire (KB32507) or even-fire (KB-32508). On even-fire motors, you can also use a stock Delco HEI (or replacement) cap and rotor from a '77-78 Chevrolet 230-250 inline six, which has an external terminal instead of a built-in coil, and add the high-energy coil of your choice.

However, the most popular ignition system among racers for GM V-6's is the addition of an MSD-6A multi-spark module from Autotronic Controls of El Paso, Texas, (part 6200—specify 4-6 cylinder unit) to the above components. This is an excellent system which works with either odd- or even-fire distributors, retaining the large cap (for reduced ionization contamination) and the big plug wires and boots. Jim Ruggles uses the setup on all his IMSA racing motors (using the Chevy six cap and rotor with a Mallory coil) and says, "It's an absolutely bulletproof ignition—and it's cheap."

Kenne-Bell markets a complete MSD conversion kit for even-fire (KB43510 and odd-fire (KB32509) Buick HEI distributors using a conversion dust cover and MSD coil. The new Accel Laser I system is much the same in function and installation to the MSD, except that it produces a longer duration spark instead of multiple sparks.

Stinger Ignitions (Midway Industries, Midway City, California), highly respected by drag racers, offers Magna Pulse (street performance), Stinger I (race version), or S-4 (race version with high-rpm retard) ignition systems to adapt to the stock HEI distributors in even- or odd-fire Buicks. However Bill Hays of Stinger feels that the physical size and weight of the HEI distributor components (especially the rotor) are a detriment to high-rpm ignition stability, and the problem is compounded by the V-6's inherent vibration characteristics. Therefore Stinger has just introduced all-new electronic ignition systems built in the early-style, small-diameter Delco distributor body for the even- and odd-fire Buick V-6's. These systems, including a complete distributor, will be available in Magna Pulse, Stinger, S-4, or Rapid Fire configurations. As for ionization or cross-fire problems with a smaller distributor cap, Hays replies, "only multiple spark, capacitive discharge ignitions create cross-fire problems. Continuous-spark, non C-D systems, such as ours, work fine in a smaller distributor."

Finally, a tried and proven ignition especially popular with drag racers is the simple, self-contained Vertex magneto. Although they are now quite expensive, a six-cylinder model is available to fit the Buick (in even-fire pattern only) from Joe Hunt Magnetos (Torrance, California) and other Vertex dealers.

HEADERS

You'll have little problem finding ready-made headers for most Buick V-6 installations. Since all Buick V-6's share the same exhaust manifold bolt pattern and relative port shape, either headers or stock exhaust manifolds will fit universally. However, the exhaust ports are raised nearly 1.0-inch on '79 and later heads, which means

Kenne-Bell initially ran this scratch-built tunnel-ram on their gasser. By turning the carbs at an angle and blocking off one throttle bore in each, they arrived at a workable "six-barrel" combination.

headers will likewise sit higher in the chassis, possibly causing some clearance problems if they are designed for earlier heads. Another consideration for header clearance on V-6's, especially odd-fire motors, is the horizontal rocking inherent to these engines which, accompanied by the "loose" factory motor mounts, requires more than average lateral clearance between headers and chassis components—not to mention a sturdy yet "shakable" exhaust system.

Several companies, such as Hooker, offer a good selection of individual-tube headers to fit Buicks in a variety of chassis and body combinations. Hooker has given special attention to the Buick V-6, and they make headers to fit most popular factory V-6 installations, plus several V-6 swap combinations including Vega, LUV, Camaro, Nova, and Chevelle bodies. Typical Hooker Buick V-6 headers have 1.5-inch diameter primaries which average 28-30 inches long, and 2.5-inch diameter, 8-inch long collectors.

After extensive testing with the

Buick, Hooker's Jack Davis discovered that the best "open" header designs did not work well when mufflers and tail pipes were added. On the other hand, designs which showed the best gains through the mufflers were just slightly better uncapped. The Hooker "ready-made" headers favor the better-through-the-muffler configuration, since this is the only design that gives improved performance both ways, and they realized that the vast majority of headers sold would be used on the street, with mufflers attached, most of the time. This is highly commendable, since far too many headers are designed on the dyno, or on race cars (if they are tested at all, other than for chassis fit), with little thought to their performance on the street.

Now let's look at some other header styles for the Buick V-6. In early tests, Smokey Yunick fabricated a complex 6-into-1 header for his Buick dyno engine. Using 30-inch primaries dumping into a single 4-inch collector at the back of the motor, it was reportedly good for 25 extra horsepower over two 3-into-1 headers. Obviously such a design would be difficult to fit into most cars, but this "360-degree" header system is being used on some circle-track cars, such as Ray Baker's ASA stocker. One thing is for sure—a V-6 using these headers sounds beautiful!

Smokey's "practical" header design is also unique. It features three short (15-inch) primaries on each side joined together at the exhaust pipe flange, with virtually no collector at all. These "Smokey" headers are now being marketed by Weiand in one "universal" style. Their small size and shape allows them to fit many cars and makes them practical for sideways-mounted Buicks in X-bodies. However, they are not specifically designed to fit any one V-6/chassis combination, and therefore might cause clearance problems in any particular car. To run

"uncorked," they will definitely require a length of header pipe of some sort to direct exhaust, heat, or flames out of the engine compartment.

Kenne-Bell concurs that the V-6 likes relatively short length, large diameter primaries. They now offer a series of "3-in-1" headers for the Buick to fit a variety of popular body styles, including the Skyhawk and Jeeps. These headers feature short 1-5/8-inch diameter primaries blending into a long 2.5-inch diameter collector which has a slip joint to allow adjustment of total collector length. In drag tests against headers with three long 1.5-inch primaries (which K-B formerly made), the new "3-in-1" style picked up 1.52 miles per hour and trimmed almost .20 seconds off the E.T. These tests were made without mufflers.

Street rod builders who are using the Buick in an early Ford or other similar chassis will be interested in a close-fitting, nicely-shaped set of headers made for this purpose by Sanderson Headers (South San Francisco, California) and marketed by Total Performance (Wallingford, Connecticut). To give you some idea how they might fit a street rod chassis, these headers extend 4 inches laterally (to the outside of the pipe) from the head; the distance from the upper flange bolts vertically to the top of the collector is 14.5 inches and 19 to the bottom of the collector; and the collector extends approximately 11 inch-

Mallory makes several distributors and ignition systems for the Buick, including the rugged dual-point "YL" model. It is very compact, providing extra clearance for blowers or other custom induction.

HEI distributors were used on both even- and odd-fire motors. The odd-fire uses a V-8 cap with two blanked terminals (shown on left). On the even-fire cap the terminals aren't evenly spaced, but the pattern is symmetrical. Both use wide terminals on the inside of the cap to extend the contact time with the rotor.

Another difference between even- and odd-fire engines is the front cover. The odd-fire has a cast timing marker (left), while the even-fire has a bolt-on indicator. Timing marks vary on both the cover and the crank pulley of even- and odd-fire motors—if you mix components you'll get some strange timing readings.

Accel makes a BEI breakerless distributor to fit the even-fire Buick, and it can be plugged into the new Laser I CD system.

es straight back from the rear primary tube. Configuration on both sides is the same.

SUPERCHARGERS AND NITROUS-OXIDE

Considerable coverage had been given in Hot Rod Magazine (May, June 1980) to a small, new, Roots-type supercharger and intake manifold made specifically for the Buick V-6 by Dyer's Machine Service (Summit, Illinois). However, this blower, installed on the Hot Rod project '32 Ford roadster, was a prototype. The unit, labeled the DMS-1, was still not in production as of this book's publi-

cation, but it should be on the market some time in the summer of 1982. It is designed as a low-profile unit, with a short, open manifold and a carburetor mounting base machined into the top of the blower case. The small supercharger is designed to be overdriven 25-30% to put out 7 to 8 pounds boost (on the Hot Rod test car, the prototype blower was overdriven 41% to produce 7 pounds of boost by 5600rpm). It is scheduled to be sold as a complete kit only, with manifold, blower, drive assembly, a re-jetted 625cfm, Carter AFB 4-barrel, and fuel line.

Although no other blower manifolds are currently made for the Buick, Blower Drive Service (Whittier, Cal-

ifornia) markets blower kits for the Buick using either a GMC 4-71 or a Magnacharger adapted to an aftermarket 4-barrel intake manifold. Their kits include blower, modified manifold, drive assembly, and carburetor adapter. The Weiand street 4-barrel intake manifold, because it comes with a bolt-on carburetor plate, is fairly easy to convert to a supercharger mounting plate. If you have machining talent and facilities, it wouldn't be difficult to adapt a GMC or Magnuson blower to this engine.

Nitrous-oxide injection, like a supercharger, works the same on any type of engine. If you use an under-the-carb adapter plate to feed the nitrous-oxide and extra gasoline to the motor, nitrous is by far the simplest and most literal form of bolt-on horsepower you can buy. Today's half dozen or so nitrous kit manufacturers have the system "dialed in." Properly installed, a moderate boost, carb-plate nitrous injector will provide strong, instantaneous, safe horsepower for short bursts of speed. It can be used on otherwise-stock engines, and has the added advantage of being a knock suppressant itself, so you don't need to add water injection or gasoline

Many racers swear by the Autotronic Controls MSD-6A (or 6T, for circle-track) multi-spark systems, which adapt to the stock HEI distributor. Shown is a complete conversion kit for the Buick, available from Kenne-Bell.

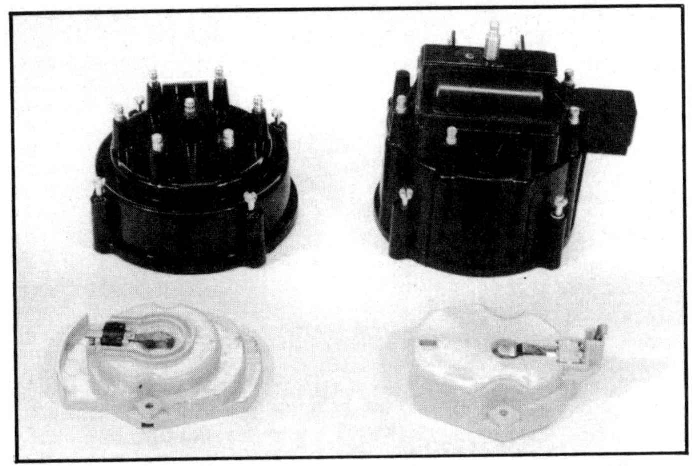

Several ignition companies now offer kits which adapt to the GM HEI distributor, replacing the stock coil with a "conversion" dust cover (right) and an external coil. A neat trick, which lowers the profile of the big HEI, is to use a Chevy straight-six cap and rotor (left).

Stinger Ignitions, which also offers conversion kits for the HEI, has just recently introduced new small body breakerless distributors for both the odd-fire (right) and the even-fire (left) Buicks. These can be plugged into any of the Stinger street or competition ignition systems, such as the Stinger I shown above.

additives along with it (as you often do with blowers or turbos).

The only disadvantages to nitrous are (1) that the system does absolutely nothing for engine performance during the vast majority of driving time (some may view this as an advantage, as it may well be), and (2) that the nitrous bottle must be regularly refilled, which is both bothersome and adds to the overall cost of the system. Nitrous-oxide injection is obviously most practical for drag racing and street machines.

Any of the nitrous manufacturers can put together kits to fit under the carburetor on Buick V-6's (though the late spread-bore 2-barrel may require a custom-made plate), or they can plumb a stock or aftermarket intake for under-the-manifold, direct-port injection. Kenne-Bell lists nitrous carb adapter kits to fit pre-'79 Rochester 2-barrels, Holley and AFB 4-barrels, and '79-up Rochester spread-bore 2-barrels. Nitrous Oxide Systems (Long Beach, California) also recently displayed a simple two-nozzle system for adding nitrous to factory turbocharged motors.

THE TURBOCHARGED BUICK

FACTORY TURBO SYSTEMS

Although Chevrolet and Oldsmobile dabbled in factory turbocharging with their early 60's Corvairs and F-85's, Buick since just 1978 has put more turbocharged vehicles on the road than any other automobile manufacturer—ever. In fact, they have very likely built more turbocharged en-

gines than any one aftermarket turbocharger company, if not all combined. This means, if you follow our drift, that the best source for turbocharging a Buick...is Buick.

Turbocharged V-6's first appeared in 1978 in Century/Regals, Rivieras, and LeSabres. Initial units came with a 2-barrel carburetor (the spread-bore Rochester E2ME). Less than 2000 of these versions were built before Buick switched to the Rochester Quadrajet 4-barrel for all subsequent turbo models. On Rivieras, the turbocharger and carburetor mount at the rear of the engine, over the bellhousing, for hood clearance. On other models, the turbo and carb sit just above the rear of the manifold. Both are compact units, though the Riviera style would interfere with the firewall in most other cars (except X-bodies). All Buick turbos are draw-through systems, using AirResearch T3 turbo housings, blowing into a special aluminum intake manifold with a small round inlet attached to the turbo by a cast elbow.

Although the basic components of the Buick turbo system have remained the same since its inception, there have been some "tuning" changes along the line. The initial concept favored maximization of fuel efficiency. The turbo system was controlled by an electronic "brain" module which included an engine knock sensor to

retard ignition timing any time the motor started to ping. The turbo was limited to nine pounds of boost maximum by a wastegate operated by a small vacuum canister. Other controls included limitation of fresh air intake to the air cleaner, carb choke setting, and so on. In 1982, however, Buick decided to retailor the turbo package to a more performance-oriented mode. Besides changing the "trim" of the turbine blades in the T3 blower for quicker response, they also eliminated exhaust-heat cross-over passages from the intake manifold entirely (the carb is given some initial warm-up from the EGR system instead), an after-the-catalytic-converter dual exhaust system became standard on all models (it was previously optional on some), the axle ratio was dropped to 3.08, and the torque converter stall speed was raised a little to allow more turbo boost build-up on take off. These changes, as a package, clipped a full two seconds off 0-to-60 acceleration times for the new turbo models, according to Buick. So far, only 231-inch versions of the Buick have received the factory turbo system.

Obviously, you can buy the turbocharging pieces from Buick and screw them onto any Buick V-6. Far better yet, you ought to be able to find a decent supply of these turbo charged motors in the wrecking yards by now, from

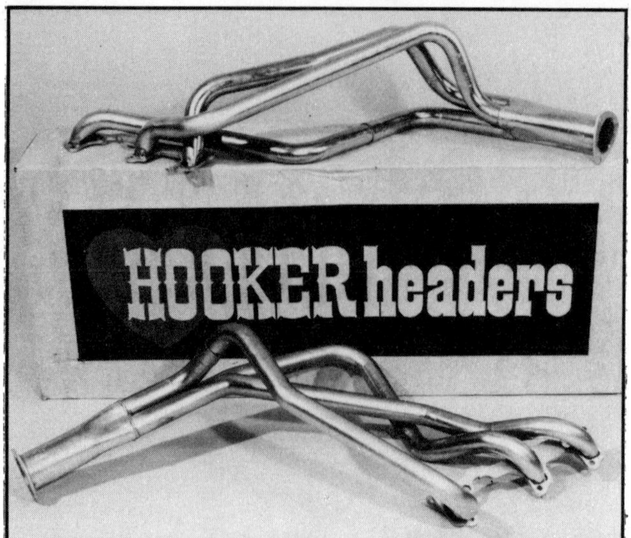

Typical Hooker street-type headers for the Buick V-6 include this version for swapping the motor into a '70-74 Toyota pickup.

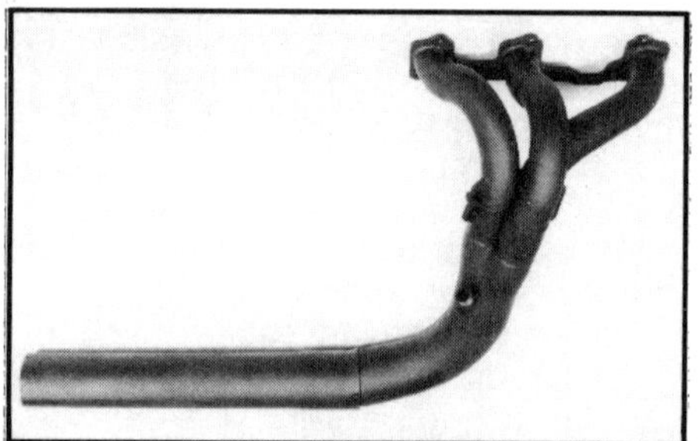

Hooker Headers made this set of custom, pro-style headers for the V-6 in Kenne-Bell's drag Skyhawk. Note the short primaries.

You can also get a Vertex magneto to fit the even-fire Buick, as seen here on a Smokey Yunick dyno motor. Note the brace—vibrations in the V-6 can cause problems with accessories like this heavy magneto.

which you could purchase the necessary parts (intake and exhaust manifolds, crossover pipes, turbo and carburetor assembly, and wastegate). Can these components be added to an otherwise stock V-6 without further modification?

The only two internal differences between production turbo motors and the naturally-aspirated variety are the considerably stronger rolled-fillet crankshaft and the tougher version of the cast piston. The turbo pistons look similar to other stockers and have the same 8.0:1 compression ratio, but they have an extra strut cast on the inside and both the ring grooves and the piston barrel are cut differently to allow for greater heat expansion.

Adding a modest (7- to 9-pound boost) turbo system, whether factory or aftermarket, to a V-6 equipped with standard crank and pistons would probably cause no problems. The factory likes to build in a wide margin of safety, and the turbo motor can take quite a bit more punishment than the stock blower puts out. Turbochargers (as well as superchargers and, to a lesser degree, nitrous-oxide injection) do significantly increase both heat and pressure in the cylinder, however, which in turn promotes detonation, especially with today's gasoline. The pistons are the components most vulnerable to these excesses. So, to build an extra margin of safety into your motor if you are adding a turbocharger to it, it would be smart to add a set of the relatively inexpensive TRW forged 8:1 replacement pistons. Although they aren't racing pistons, they are designed specifically for high-heat, heavy-load conditions and they are tougher than the factory cast turbo

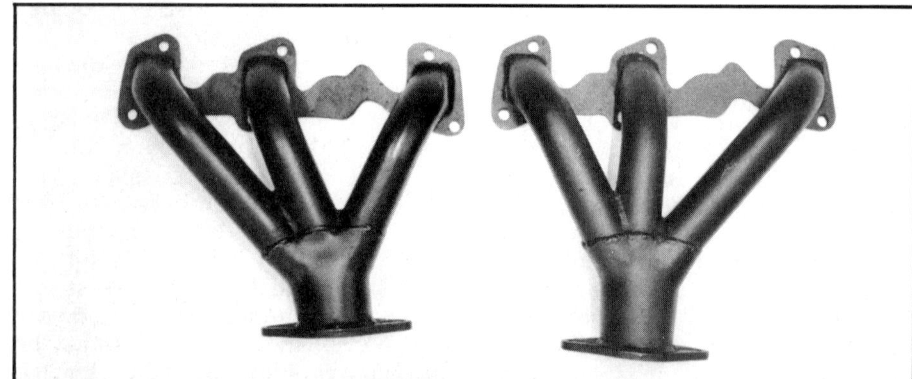

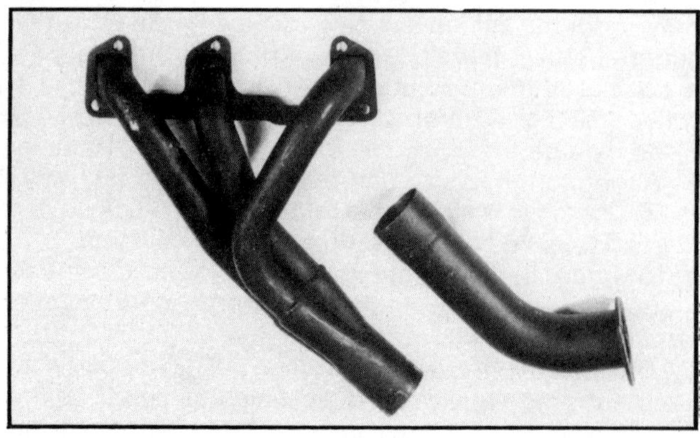

Weiand's "Smokey" headers for the Buick are short and to the point. They aren't made to fit any chassis in particular, but are small enough to squeeze into almost any chassis.

Kenne-Bell's new "3-in-1" headers for the Buick fit tight to the motor and feature an adjustable, slip-fit collector.

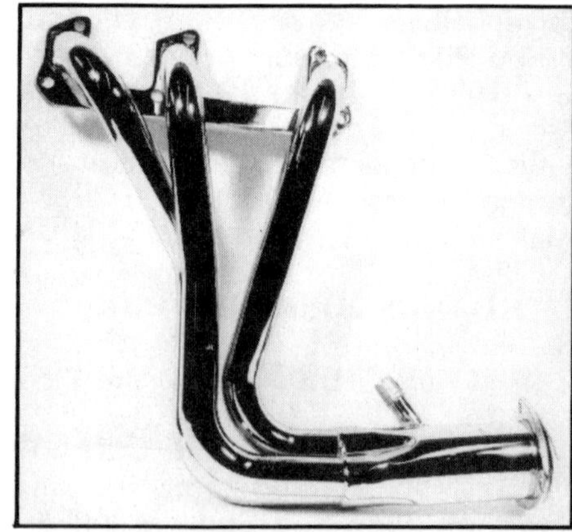

These good-looking Buick headers were created by Sanderson for the blown Buick V-6 in Hot Rod magazine's '32 Ford roadster. They are available from Total Performance.

This is the prototype Dyer blower made specifically for the Buick V-6. Note the low profile of the overall installation.

pistons. (The forged TRW's are actually cheaper than the cast factory pistons, retail.) Besides lower production cost, the factory stuck with a cast piston primarily to reduce engine noise. Forged pistons, which expand more when they heat up, must be installed at looser clearances, and they may cause some "rattle" on cold starts, but that shouldn't bother a performance driver.

AFTERMARKET TURBOS

Besides factory components, a wide variety of turbochargers and universal adapting pieces is available

from numerous companies like Spearco and Ak Miller. You can readily design and fabricate your own turbo system, or have one custom tailored to your engine and chassis to produce mild to wild boost pressure. The fact that cast-iron turbo outlet exhaust manifolds are available from the factory makes fabrication of such a system relatively easy. You could also use the factory turbo intake manifold, but a stock 2- or 4-barrel intake would work just as well, if not better (the early 2-barrel intake is preferable to the later spread-bore manifold—but use a man-

Blower Drive Service 4-71 GMC kit for the Buick uses a cast aluminum blower mounting plate welded into an aftermarket manifold.

Adapting the Weiand manifold to accept the blower was relatively straightforward. The manifold makes this job easy because it is designed to accept a bolt-on carburetor plate. Jerry welded the EGR passage at the rear of the plenum.

Jerry Magnuson is shown setting up one of his Magnachargers with dual Weber 2-barrels on Roy Fjastad's Buick V-6 (shown in completed form on the cover). The blower was easily adapted to the engine by using a Weiand street 4-barrel manifold, removing the carb plate and machining a blower mounting plate to bolt onto it.

ifold to match the heads on your motor). It would also be wise to use the '80 turbo intake manifold gasket to reduce exhaust-heat cross-over.

The only company currently making a bolt-on aftermarket turbocharger kit for the Buick V-6 is Kenne-Bell. This draw-through system, which was designed before Buick brought out the factory turbo, adapts to a stock 2-barrel intake manifold, uses a fabricated right-side header, and is available with either an SU single-throat or Dellorto dual-throat, side-draft carburetor. A non-adjustable system which does not use a wastegate, it can be ordered in 7-, 9-, or 15-pound boost versions. K-B also offers individual turbocharging components including water-heated adapters for SU, Dellorto/Weber, or 4-barrel carburetors which will fit the factory turbo, and an advance/retard vacuum canister to cut back ignition lead under boost.

A considerably more radical (and expensive) dual turbo setup has been adapted to several Buick V-6's by Gale Banks Engineering (San Gabriel, California). These systems are generally custom tailored for specific use and installation, but are of a blow-through design with the turbos feeding into a pressure box atop a 4-barrel carburetor on a performance manifold. Gale has built such systems for both street and racing, with the turbos mounted either at the front or the rear of the engine. Banks has also been working with the factory "knock sensor" ignition control system, adapting it to retard spark timing in a mode compatible with the performance turbo system, rather than using water injection or a vacuum-type advance/retard distributor canister.

MODIFYING THE STOCK TURBO

Admittedly, the stock Buick turbo-

The blower bolts to an adapter plate that fits on the Weiand manifold, in place of the stock carburetor plate. The adapter plate is machined from heat-treated .500-inch aluminum plate. Note that top of the plate has been machined for clearance with the blower rotors.

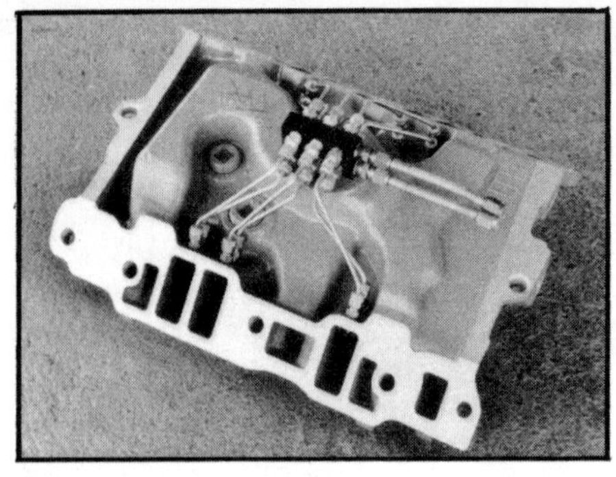

A typical direct-port Nitrous Oxide injection system has been plumbed into the underside of this factory, aluminum, 4-barrel 4.1-liter manifold.

Nitrous Oxide Systems of Long Beach, California, tapped a pair of nozzles into the carburetor housing—just ahead of the turbo—on this 1982 factory blower setup to add a shot of laughing gas to the boost.

charger system is no barnstormer. But it does contain all the same essential ingredients as most high-performance turbo setups—plus several extra valves, wires, hoses, and special linkages to placate both the government and the company execs who write warranty policies. Although the latest (1980 California and '81 all-states) Buicks are all now fitted with a comprehensive, and complicated, Computer Control System (see the 60-degree Chevy section for more details), even these, and especially earlier, turbocharged V-6's are far from "tamper-proof." The same turbo that is limited to less than nine pounds boost in factory form can be "tweaked" to produce nearly twice the pressure just by turning a few screws...more or less.

If you are really serious about turning up the boost on a production turbocharged Buick, we'd recommend that you begin inside the motor. Although the very same turbo unit can be exceptionally docile in stock, restricted form, it is fully capable of putting out 15-20 pounds of boost completely unfettered. That sort of pressure, especially if applied under rigorous or racing situations (and that's what you have in mind, right?) will affect engine components the same way whether it's coming from a "production" Buick turbo or a GMC supercharger. If you

are going to treat your V-6 like a high-performance engine, you had better build it like one.

The first area of concern for a hard-running turbo motor is the pistons. For high-rpm/high-boost or sustained boost situations you should use a good quality forged racing piston in a compression ratio of 8:1 or less. Such pistons are available on a custom-order basis from manufacturers such as Arias, Venolia, Forgedtrue, etc., and they can be ordered for the Buick V-6 through Kenne-Bell. A second area of major concern in the Buick V-6 when higher cylinder heat and pressure are applied by a turbo is adequate cylinder head-to-block sealing. Peculiarities of the 8-bolt, easily warpable Buick head design are discussed in the head and block sections, with recommended cures which apply especially to turbocharged motors. Third, although more fuss has been made over Buick V-6 oiling than is perhaps warranted, remember that a turbocharger must be fed a constant supply of clean, pressurized oil from the engine oil system. This means a slight addition to the total

This 1978 Century has plenty of changes on the outside, but the engine remains a stock factory turbocharged 3.8-liter. It's hard to find the turbo amid the accessories, but it's there.

volume and pressure drop in the system, plus a greater need for good filtering. Fourth, any high-performance turbocharged motor should be fitted with top quality rings and bearings and expertly blueprinted to performance specs. The rolled-fillet crankshaft that comes stock in all turbo motors should handle plenty of abuse, but the strength of both it and the stock connecting rods can be significantly increased by a heat-treating process such as Tufftriding.

The scope of this book does not allow for a full discussion of either turbo engine building practice or the subtleties of turbocharger system design and tuning. The point is that you can build a real high-performance or competition turbo system around the basic production Buick parts, but it will require substantial modification to the engine, as well as to the turbo system, plus a good working knowledge of turbos on your part.

The above having been said, what about the possibility of turning up the performance of the factory turbo system in your daily driver without having to rebuild or throw away half of the factory components? It certainly can be done, and relatively easily, as outlined in past magazine articles (specifically *Hot Rod Magazine*, July '79 and June '81). Maximum boost on

With some of the clutter stripped away, it's easier to see the layout of the factory turbo setup. This is the Riviera system with the blower mounted low and to the rear of the manifold.

factory turbos is limited by a wastegate mounted on the exhaust turbine. This valve "bleeds" incoming exhaust pressure past the turbocharger when a certain pressure is reached in the intake manifold. It is controlled by a diaphragm canister mounted on the intake turbine housing and an actuating rod connected to the wastegate. A very simple way to change the maximum boost delivered by the stock turbocharger is to alter the length of the actuating rod to make the wastegate open at a higher manifold pressure. The replacement Buick wastegate actuator canister (part 1261048) actually comes with an adjustable rod, but it is simpler (and much cheaper) to cut and thread the rod yourself and add an internally-threaded sleeve to the clevis end (you should be able to find suitable pieces from carburetor throttle linkage kits).

On 1981 turbo models only, which use a "normally-open" wastegate, the length of the actuator rod must be increased to produce more boost (1/8- to 3/16-inch will bring boost up to 12-14 pounds). On all other models, shorten the rod to increase boost (about .50-inch for 15 pounds). Before you start recalibrating the wastegate, however, install a good manifold vacuum/boost gauge in the car. Then adjust the actuator rod in *small* increments, performing extensive driving tests between each change to see how much boost the system is delivering. Otherwise you might end up with vital engine components lying on the highway.

Another way to modify boost pressure on the stock turbo setup is to install an aftermarket, adjustable wastegate controlling mechanism. The Dial-A-Boost (manufactured by Turbo Tom's, Atlanta, Georgia) is such a device, which can be hooked into the Buick turbo system to allow remote adjustment of the boost pressure, even from the passenger compartment. This unit sells for about $25. Turbo Tom's also sells turbo-compatible water-injection systems.

However, just turning up the boost on your factory turbo isn't going to do much for the engine other than make it "knock." Engine knock (detonation) will not only limit performance (if not break pistons), but will also trigger the electronic knock sensor to drastically retard ignition timing, causing the engine to fall flat on its face. The solution is to take several steps to reduce the knock tendencies of the motor.

First of all, naturally, use the highest octane gasoline available. Secondly, on '78-79 models it would help to install the '80 intake-manifold gasket to reduce manifold heating. On water-heated carburetor plenums, you can install a small shut-off valve in the water line to reduce flow in winter, and shut it off completely in summer. And you can change the stock 195-degree engine thermostat for a 180-degree unit.

It would also be wise to install a water-injection system on the engine.

If you don't have a factory turbocharged Buick, you can always build your own. Kenne-Bell offers the only complete turbo kits for Buick V-6's, designed to adapt to a stock 2-barrel manifold. The completed installation (above) has an AFB 4-barrel while the kit (left) uses an SU side-draft carb.

These are available from a variety of aftermarket sources, but most are designed to activate when the manifold vacuum decreases to a certain level. Such water injectors will not work effectively with the higher manifold pressures generally found in turbo motors (unless the control module can be modified or adjusted to accommodate "boost" conditions). If you can't locate a unit that is compatible with a blown motor, it may be possible to wire the water-triggering module into the pressure-actuated switch (mounted on the inner fender panel) which controls the yellow or orange boost indicator lights in the Buick dash panel.

Each of these steps will help reduce detonation in the engine under increased boost, which will in turn keep the knock sensor from retarding ignition timing. Since we are talking about an otherwise-stock motor, we would suggest you leave the knock sensor as is to save engine internals from detonation damage if you get carried away with increased boost or overly aggressive driving.

The above minor modifications will allow you to increase boost to a level of 12 to 14psi, but don't be surprised if that still doesn't turn your Buick into a drag-strip terror. If the motor resides in a Regal, Century, or Riviera, you are lugging around 3500-plus pounds of car with a 231-inch engine (which

If you really want to crank some pressure into your V-6, try a Gale Banks dual-turbo system. This is a blow-through design feeding into a 4-barrel carb on an after-market manifold.

If you have a complete machine shop at your disposal, you can really get carried away and create something like this around your Buick V-6. Hand-crafted work of art, using two small Japanese turbos and two Weber side drafts, will power Jim Ewing's new '34 Ford street rod.

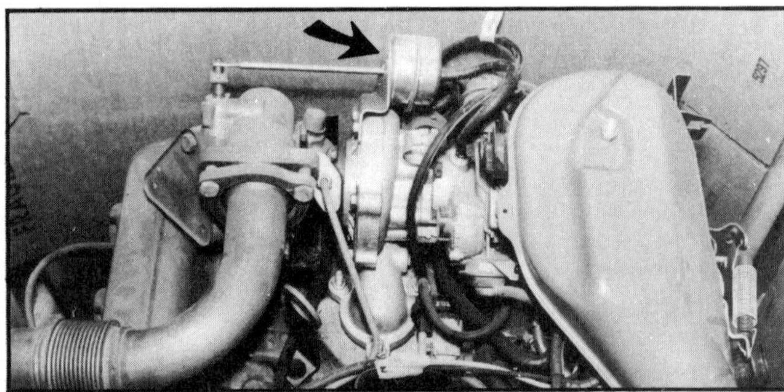

The diaphragm and control rod (arrow) which operates the wastegate on the factory system is easy to locate and easy to adjust or modify to increase the effective boost. This is a Century/Regal system; other setups might mount the cannister in a different location.

engine. The flywheel may help the motor run smoother on the highway, but it doesn't let it rev quickly for drag racing, plus it is so thick that only a thin, diaphragm-type, factory pressure plate will fit between it and the bellhousing.

Both Kenne-Bell and McLeod make steel flywheels to fit the V-6 which weigh 30 to 35 pounds and will accept 9.125- (stock size) to 10.5-inch pressure plates. Although lighter (15 to 18 pound) aluminum flywheels are available from the same sources, Jim Bell states that for drag racing the 35-pound wheel is essential to help launch most V-6-powered vehicles. A light wheel should only be used for

further adds to detonation problems). If the vehicle is a pre-'82 model with 2.70 differential gears and an automatic transmission, what it really needs is lower gearing and a looser torque converter, plus a less resrictive exhaust system. These elements are standard, as we mentioned, on '82 and later turbo cars and optional on certain '81's (3.08 gears, higher stall converter, dual exhaust). Adding them to an earlier model falls outside the realm of "minor adjustments," to be sure, but they are really necessary to allow the engine to build rpm in lower speed ranges, which in turn allows the turbo to create effective boost and get a heavier car moving.

With all the above changes to a factory system you can expect quarter-mile improvements in the range of 1 to 2 seconds and 5 to 10 mph, plus substantially more power at freeway speeds. Further improvements can be gained by some reworking of the carburetor (if it's not computer-controlled), especially to increase the rate of opening on the secondaries. *However, remember that turning up the boost on an otherwise stock V-6 can surely lead to parts breakage if carried beyond practical limits—and such fail-* *ures are not covered in your new car warranty.*

BELLHOUSING, FLYWHEEL, CLUTCH

The early 198-inch '61-63 Buick V-6 used an unusual bellhousing pattern and came with 3- or 4-speed stick transmissions or a Dual-Path 2-speed automatic. If for some reason you are using this block and want to adapt it to later GM automatic or standard transmissions, special adapter plates and flywheels for this swap are available from D&D Fabrications (Box 55, Rochester, Michigan 48063). All '64 and later Buicks use the standard Buick, Cad, Olds bellhousing pattern which will accept Turbo 350 and 400 automatics, 3-, 4-, or 5-speed sticks, or other GM late-model transmissions.

Since the vast majority of Buick V-6's have the current bellhousing pattern, fitting a good automatic to these engines is simple—take your choice and bolt it up. With standard transmissions things get a little stickier, though. To begin with, odd-fire motors use a very thick, heavy flywheel (approximately 50 pounds) to help smooth out the firing pulses of the

On this stock Riviera turbo motor, swapped into an '80 Skylark X-body, you will note that the wastegate control canister (arrow) is just ahead of the turbo housing. It has been connected to a variable boost control unit which is set to deliver 15 pounds of pressure. Besides tuning changes, the only other modification to the package is a boost-compatible water-injection unit.

oval track.

As we mentioned in the crankshaft section, factory flywheels (or flexplates) contain part of the crankshaft balance—this condition is known as "external balance" or "Detroit balance"—and the balance is slightly different for even- and odd-fire motors. Consequently, production flywheels or flexplates should not be swapped from one type of motor to the other.

Most aftermarket flywheels for the Buick V-6, however, have the same "Detroit balance" built in for use on either type of engine. Many purchasers bolt them on and use them as-is, but they are intended to be rebalanced along with the crankshaft during a performance engine blueprint. Another alternative, as we mentioned, is to have a balance shop match the new flywheel to the one you take off the engine. If the shop doesn't have a special fixture for this process (few do), the two flywheels can be spun together, indexed 180 degrees apart, and the new one can be drilled until the combination has zero unbalance.

Two different styles of bellhousing and clutch linkage have also been used on Buick V-6's. Regals and Centuries have a horizontal clutch-throwout arm, like a Chevy, and use mechanical linkage. Skyhawks and other GM H-bodies, however, have the clutch arm angled at approximately 45 degrees from horizontal, towards the bottom of the car, and use a cable release mechanism.

Steel one-piece bellhousing/scattershields have long been available for the horizontal-arm, mechanical-linkage setup (Lakewood part 15120 or Kenne-Bell part KB16701), or Chevytype scattershields can be used by drilling four holes in the flange to fit the Buick. Until recently nothing was available for Skyhawk installations, but Lakewood has just released a new bellhousing, part 15121, with the angled clutch arm. It hooks up to the stock clutch cable and uses the stock pivot, bearing, and throwout arm.

McLeod offers a wide variety of

Pre-'64 Buicks used this odd-shaped bellhousing, identical to that on the 215 V-8. Popular GM standard transmissions will fit it, but modern automatics won't bolt to the block.

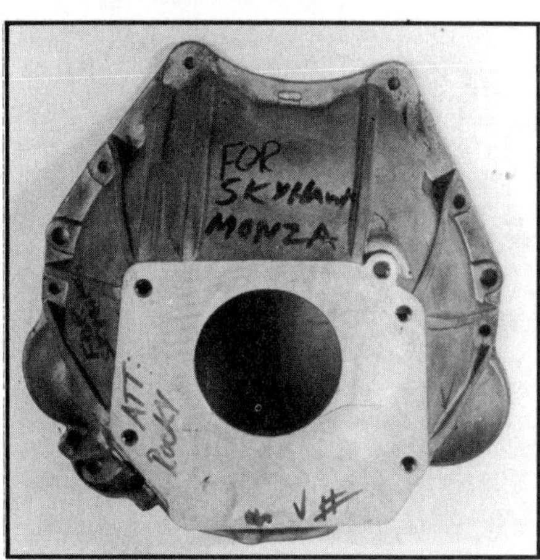

The '64 and up Buicks use the current Buick, Olds, and Cadillac bellhousing pattern. Skyhawks used a cable-release clutch fork positioned on the lower left.

pressure plates and complete units to fit their flywheel (which is drilled to accept 9.125-, 10-, or 10.5-inch pressure plates). However, the stock cable-actuated mechanism does not provide enough "throw" to operate the popular Borg and Beckstyle pressure plate, so they recommend using a "combo" Borg and Beck/Long-style 10.5-inch plate for racing, or their Chevy-style 10.5-inch diaphragm for street.

BUICK V-6 ENGINE SWAPS AND KITS

A couple of magazine articles have been published showing the transplant of Buick V-6's into X-body cars (*Hot Rod Magazine* October '79 and December '80), producing a light, quick combination with the larger Buick motor. Now that the 3.0-liter Buick 90-degree motor is available in the front-wheel-drive chassis, the swap may seem academic. However, the 3-liter is hardly bigger than the 2.8-liter Chevy V-6 previously available. Mid-14-second quarter-mile times reported for one swap were recorded with a 4.1-liter Buick. This engine was mated to the X-car 4-speed trans-axle

(which has a different bolt pattern than the Buick) with a flat steel adapter plate made by Doug Roe. Since the X-car bellhousing is quite small, a special flywheel has to be found to match the X-car clutch. McLeod used a flywheel from a 273-inch Mopar V-8 for this combination, drilling it with a center bolt pattern to fit the Buick crank and an outer pattern to accept the stock X-car pressure plate and clutch, which has a 9.125-inch diameter. A Mopar starter also had to be adapted to the combination.

THE 3.0-LITER ENGINE

Of course, you can drop the 3.0-liter Buick into any X-car, since it has a bellhousing flange to match the X-car pattern. Further, since the 3.0 has the same bore size as the 3.8-liter motor, you could swap in the bigger crank, rods, and pistons to make a 3.8 with the

Although McLeod flywheels are drilled to accept a variety of clutch and pressure plate types, they recommend this 10.5-inch Borg-Beck style combination (#360103) for street and strip performance, or else the 10.5-inch "combo" Borg-Beck/Long pressure plate for cars with cable clutches.

Early Buicks used a flat flywheel (left), while '75 and later V-6's have a recessed face (right). These are performance versions of each made by McLeod.

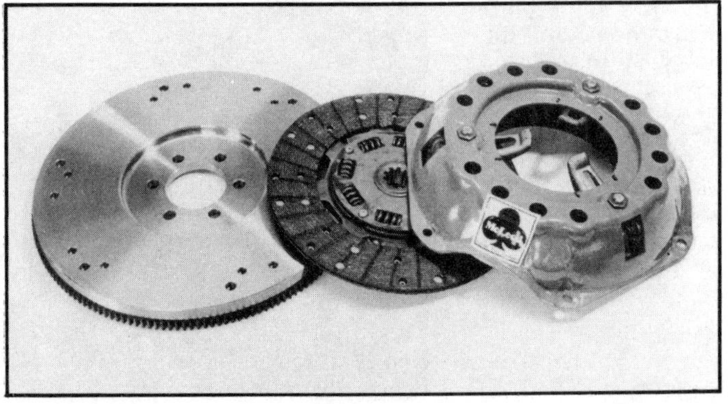

Doug Roe swapped a 4.1-liter Buick into this 1980 Skylark with surprising results. To mate the engine to the X-car transaxle, he had to make a 3/4-inch thick steel adapter plate, visible in the engine compartment photo, plus adapt the small flywheel and clutch. The motor is modified with the Smokey "5-Step Plan," producing mid-14-second quarter-mile times in street trim.

X-car bellhousing bolt pattern. The only problem is that the 3.0 crank has a significantly smaller flywheel bolt pattern and flange than the 3.8, so you cannot bolt on a matching flywheel. There are two possible solutions. You could adapt the Mopar flywheel and starter, as in the Roe swap (or purchase the flywheel from McLeod, if it becomes available), or it might be possible to turn down the flange at the rear of the 3.8 crank to the X-car diameter and drill it to accept the X-car flywheel. Buick engineers stated that they have successfully done this in the past, but they feel that the extremely small flywheel bolt pattern is marginal, at best, on a larger displacement engine, especially one to be used in a performance mode.

Since we are discussing the X-car swap for the 231, we should mention that a significantly shorter water pump is needed for this cross-wise installation. The 3.0-liter engine comes with a completely revised front cover, much shorter than other Buicks, and it will fit onto any other Buick V-6. This cover has the oil filter angled parallel to the front of the engine (this oil-filter adapter could probably be installed on

other front covers, as well) and it has no provision for a fan or a mechanical fuel pump. The X-car uses both an electric fuel pump and an electric fan. Besides allowing the 90-degree Buick V-6 to squeeze into the tight X-car chassis, this new front cover assembly makes the already-compact motor even smaller, opening up all sorts of possibilities for mini-car and street rod swaps or race car installations. If you don't want to change the whole front cover, you can shorten the front of the engine about 3/4-inch by using a water pump (with matching pulleys) from a '64-67 225. This pump is available from Buick (part 1306876), TRW (FP 1318) or Sealed Power (PL 300).

SWAP KITS

The whole idea behind the V-6 engine configuration is to make a moderate-size engine into a very compact package, so it will fit into small spaces. The Buick V-6, offering as much as 252 cubic inches, is actually shorter in both length and height than most four-cylinder engines, and it isn't much wider. Consequently, it has become popular as an easy swap into compacts like the Vega and into a

variety of mini-trucks, not to mention the X-car swap outlined above.

Several manufacturers now offer bolt-in swap kits for the Buick. Hooker Headers makes kits (including motor mounts, and transmission mounts, and headers) to swap the Buick into Chevy Vegas (part 1602), Chevy LUV pickups (part 1117), and Datsun mini-trucks (part 1116). The numbers given are actually for the headers only; the rest of the swap kit will depend on the type of transmission being used. The Vega installation uses GM motor mounts from the '75-78 Skyhawk, Starfire, or '78 Monza, which could be obtained from the wrecking yard or from any GM dealer (left frame stand, 1705321; right frame stand, 1746350; right and left U-brackets, 1250392 and 1249826; and left and right rubber mounts, 1258666 and 1255277). The Datsun installation kit (part 12616) includes engine and trans frame mounts, plus a recessed fiberglass firewall.

Herbert Automotive (5467 Ballentine, Sacramento, California 95826) makes a complete Vega/V-6 swap kit including mounts (part 27901), headers (part 27902), a modified oil pan

The new 3.0-liter Buick block is the same as the 3.8 except that the bellhousing has been changed to match the X-car transaxle. The bolt pattern on the crankshaft and flywheel has also been significantly reduced to the diameter used on other X-car engines.

(part 27906), and a special heavy-duty radiator (part 27907). Herbert also offers a V-6 LUV kit consisting of front and rear mounts only (part 23901A for '72-75, 23901B for '76-80).

The Herbert and Hooker mini-truck installation kits are for two-wheel-drive versions only. Downey Off Road Manufacturing (10023 S. Pioneer Blvd., Santa Fe Springs, California 90670) specializes in adapting the Buick V-6 to Toyota four-wheel-drive pickups. Their kit (part R-11-90) mates a V-6 with Turbo 350 automatic transmission to the Toyota transfer case using a new trans-output shaft and adapter plate. Other kit parts include front motor mounts, alternator bracket, temperature and oil sender adapters, and a flex fan. Headers, power steering bracket, and an air-conditioner pump bracket are also available.

Low manufacturing (245 W. Foothill Blvd., Monrovia, California 91016) offers all sorts of kits to put bigger engines in a variety of mini-trucks. For the Buick V-6, they make swap kits for two- or four-wheel-drive Toyotas, two- or four-wheel-drive Datsuns, and two-wheel-drive LUV's.

Finally, proving that a V-6 can be swapped into just about anything, complete kits to adapt either a '61-63 or '64-up Buick V-6 to a Volkswagen transaxle are available from Kennedy Engineered Products (10202 Glenoaks Blvd., Pacoima, California 91331). These motors have been successfully swapped into VW buses, Baja bugs, dune buggies, and VW-based kit cars. In fact, KEP highly recommends the Buick V-6 as an excellent power replacement for anemic VW buses. They also make kits to adapt early or late Buick V-6's to Porsche and Corvair transaxles, as well as a bellhousing that will mate '64-up engines to the Pantera ZF 5-speed transaxle. Incidentally, Kennedy Engineered Products makes similar kits to mate the Ford/Capri V-6, the Chevy 60-degree V-6, the Chevy 90-degree V-6, or even the Volvo, Renault, Peugeot V-6 to VW or Porsche transaxles.

The 3.0-liter also has a revised front cover which reduces the length of the engine quite a bit. It will fit on other Buick V-6's, making them even shorter for swapping into tiny engine compartments.

BUICK V-6 PRODUCTION SPECIFICATIONS

	198	225	231 odd fire	231 even fire	196	4.1L	3.0L
Displacement	198ci	225ci	231ci (3.8L)	231ci (3.8L)	196ci (3.2L)	252ci	181ci
Years produced	62-63	64-67 (Buick) 67-71 (Jeep)	75-mid 77	mid 77 and later	78 and later	80 and later	82 and later
Bore	3.625-inches	3.750-inches	3.80-inches	3.80-inches	3.50-inches	3.965-inches	3.80-inches
Stroke	3.20-inches	3.40-inches	3.40-inches	3.40-inches	3.40-inches	3.40-inches	2.66-inches
Firing order	1-6-5-4-3-2	1-6-5-4-3-2	1-6-5-4-3-2	1-6-5-4-3-2	1-6-5-4-3-2	1-6-5-4-3-2	1-6-5-4-3-2
Cylinder numbering	1,3,5 left bank 2,4,6 right bank	1,3,5 left bank 2,4,6 right bank	1,3,5 left bank 2,4,6 right bank	1,3,5 left bank 2,4,6 right bank	1,3,5 left bank 2,4,6 right bank	1,3,5 left bank 2,4,6 right bank	1,3,5 left bank 2,4,6 right bank
Compression ratio	8.8:1	9.0:1	8.0:1	7.8:1	8.0:1	8.0:1	8.45:1
Main journal diameter	2.30-inches	2.50-inches	2.50-inches	2.50-inches	2.50-inches	2.50-inches	2.50-inches
Rod journal diameter	2.00-inches	2.00-inches	2.00-inches	2.25-inches	2.25-inches	2.25-inches	2.25-inches
Rod journal offset (split)	Zero	Zero	Zero	30°	30°	30°	30°
Connecting rod length	5.86-inches	5.86-inches	5.86-inches	5.96-inches	5.96-inches	5.96-inches	6.33-inches
Valve head diameter	Int: 1.500-inches Exh: 1.3125-inches	Int: 1.625-inches Exh: 1.380-inches	Int: 1.625-inches Exh: 1.425-inches	Int: 1.71 (for 79 up) Exh: 1.50 (for 79 up)	Not available	Int: 1.710-inches Exh: 1.500-inches	Not available
Valve clearance	Zero	Zero	Zero	Zero	Zero	Zero	Zero
Rockerarm ratio	1.6:1	1.6:1	1.55:1	1.55:1	1.55:1	1.55:1	1.55:1
Main bearing clearance	.002-inches	.0005-.0020-inches	.0005-.0015-inches	.0005-.0015-inches	.0003-.0020-inches	.0003-.0020-inches	.0003-.0020-inches
Rod bearing clearance	.0015-.0020-inches	.0020-.0023-inches	.0020-.0023-inches	.0005-.0026-inches	.0005-.0026-inches	.0005-.0026-inches	.0005-.0026-inches
Piston-to-bore clearance	.0008-.0014-inches	.0008-.0014-inches	.0008-.0014-inches	.0008-.0020-inches	.0008-.0020-inches	.0008-.0020-inches	.0008-.0020-inches
Piston ring end gap	.010-.020-inches	.010-.020-inches	.013-.023-inches	.010-.020-inches	.010-.020-inches	.010-.020-inches	.010-.020-inches
	Bolt torque (lb-ft)	Bolt torque (lb-ft)	Bolt torque (lb-ft)	Bolt torque (lb-ft)	Bolt torque (lb-ft)	Bolt torque (lb-ft)	Bolt torque (lb-ft)
Main caps	100	100	100	100	100	100	100
Rod caps	40	40	40	40	40	40	40
Cylinder head	80	80	80	80	80	80	80
Intake manifold	55	55	55	55	55	55	55
Rockershafts	35	35	35	35	35	35	35
Flywheel	65	65	65	65	65	65	65
Spark plugs	20	20	20	20	20	20	20
Harmonic balancer	150	150	150	150	150	150	150

CHAPTER 5

FORD CAPRI V-6

THE FORD-CAPRI V-6

The most overlooked and unappreciated of the "American" V-6's is the Ford/Capri.

Actually this engine is not of American origin, having been designed and built by Ford of Germany in 2.3- and 2.6-liter versions for their Taunus 20M sedan back in 1964. Later, when Ford of Germany introduced the Capri, a sporty 2-door sedan, they installed the 2300cc V-6 (later enlarged to 2600cc) as an option alongside the standard four-cylinder inline. In 1972, Ford's U.S.-based Lincoln-Mercury division decided to import the Capri as a sporty "compact" addition to their line, and the European 2600 (155 cubic inch) V-6 came with it as an option.

In 1974, the German-built V-6 was enlarged to 2800cc (170 cubic inches) with an increase in bore and stroke, plus a cylinder-head change. This new version came in Mercury Capris, but was also shipped to the U.S. for installation in Ford Pintos and Mustang II's, and later in Mercury Bobcats. Finally, in 1980 the importation of this engine for American Ford products was discontinued. The company cited the comparable power of the turbocharged four-cylinder, plus a greater need for the engine in Europe as reasons for this discontinuation.

The "German" Ford V-6 used in the United States is not the same as the "Essex" V-6 built by Ford of England, though both engines are quite similar. In the 1960's the British and German divisions of Ford were separate en-

The 2600 Ford, made in Germany and imported in Mercury Capris, is characterized by heads with two exhaust outlets per side. Note rear-sump oil pan used in Capris.

tities (now merged into Ford of Europe), and each company embarked at about the same time on programs to develop new V-4 and V-6 engine families. Without consulting each other, both design groups came up with surprisingly similar packages, each including a 60-degree V-4 with a counterbalance shaft in the lower block and a V-6 with the same basic block casting but without the balance shaft. The British V-6 was built in 2.5- and 3-liter sizes. Both the German and the British V-6's feature even-firing, six-throw crankshafts with very short stroke length to reduce inertia forces (i.e., vibration) to a minimum.

Unfortunately, no components from the British engine will fit the Ger-

man engine, or vice versa. This is especially frustrating since the 3-liter British engine was used extensively in racing programs in England and Europe. Ford commissioned Cosworth to build an exotic dual overhead-cam cylinder head kit for racing versions of the British V-6. This 4-valve-per-cylinder engine was not unlike the Dino Ferrari V-6, and it acted like one. Using other trick components like a billet crank, forged rods, and Lucas fuel injection, it powered several successful Ford racing machines in England and throughout Europe. In '76 and '77 it won the British Formula 5000 championship, beating cars powered by 302 Chevy V-8's. A wide range of performance equipment, including a

The 2800 has three exhaust outlets per head. Both engines use smooth, tight-fitting exhaust manifolds which help reduce overall engine size to a minimum. Note front sump on this 2800, denoting use in a Mustang II.

In 1973 Ford of England built this turbocharged, 3-liter V-6—which produced nearly 200 horsepower in streetable form—for a Capri experimental prototype. Notice the front distributor location on the British engine.

Ford of Germany instituted a Competition Department in 1968, and built several successful racing versions of the V-6 Capri. Beating cars like BMWs, 2600 and 3-liter Capris won numerous Group 2 races and the 1971, 1972, and 1974 European Touring Car Championships. Shown is a German 2600 racing V-6 fitted with Weslake heads and Kugelfisher slide-valve fuel injection.

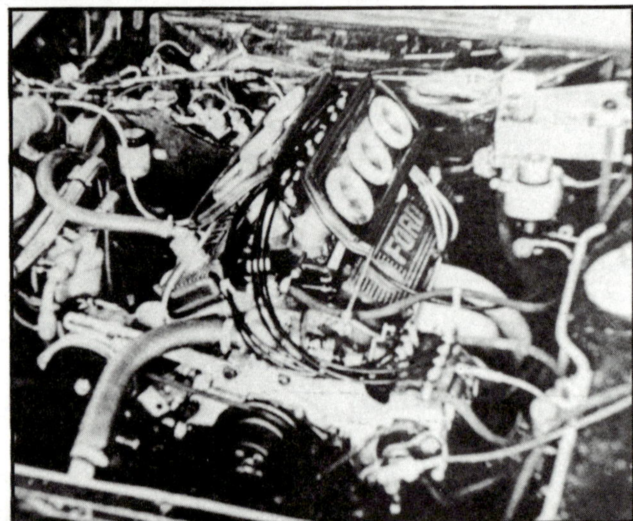

lucky to find a set these days. In 1972 the Capris, using a 2.9-liter version of the 2600 engine with Weslake heads, won the European Touring Car Championship. By 1974 the Capris were fitted with a 3.4-liter version of the British V-6 with the DOHC Cosworth heads.

Naturally such racing activity by the German Ford factory resulted in a variety of factory performance components (known as "RS" parts in Germany) which were made available through German Ford dealers. In fact, a significantly modified version of the Capri, known as the "RS 2600," was introduced in 1970 in Germany and sold as a regular showroom model. It featured a slightly longer stroke, 10:1 compression, a bigger cam, low-restriction exhaust, and a Kugelfisher "plenum" mechanical fuel-injection system similar to Rochester injection used on early Corvettes. Furthermore, the RS 2600 had stiffer, lowered suspension, Bilstein shocks, racing seats, extra instrumentation, and came with the bumpers removed. The RS model was extremely popular in Europe during its run from 1970-74, but unfortunately neither it nor any of its components was offered in the United States.

Those who wanted to upgrade their 2600 V-6 to the "RS" specs in Germany could buy a special "tuning kit" consisting of a stroked crank, heavy-duty rod bearings, high-flow oil pump, heavy-duty head gaskets, special vibration dampener, sport exhaust, and a 272-degree "RS Sport" cam. The 2.2mm stroked crank, used with stock pistons and rods, increased the piston deck height, thereby increasing the

Wade Roots-type supercharger kit, is also available in England for the 3-liter British V-6.

In 1968 Ford of Germany established a small competition department and began entering specially-prepared Taunus and Capri 2300 V-6 sedans in rallies, marathons, hillclimbs, and Group 2 Touring Car Championship races with considerable success. Weslake of England was commissioned by Ford to modify some of these engines, and in 1970 they designed and manufactured 50 sets of aluminum pushrod-type heads for this engine to be used exclusively by the factory racing team. These Weslake heads featured ports and valves similar to the famous Weslake Ford small-block V-8 heads used in Indy racing; on the V-6 they were fitted with German Kugelfisher direct-port fuel injection. These heads will also fit the 2600 Capri engine, but you would be very

Similar in appearance and principle to the Rochester fuel injection once available on Corvettes, this Kugelfisher-injected 2600 was available in RS Capris in Germany in the early 1970s. Just recently, a similar injection setup has been added to 2800s (German) now being installed in RS Capris in England.

John Norris competes very successfully in Southern California slalom and solo competition with this well-used 2600 Capri. With the stock 2-stage Weber carb, a Norris 303-S cam and lifters, TRW gears, minor head work (around valve pockets) and Thunderbird headers, Norris says the engine "makes power from 2500 to 7500rpm."

compression ratio. Other RS parts available in Germany for the 2600 included forged pistons, heavy-duty rods, stainless steel valves, heavy-duty valve springs, tube headers, a 316-degree cam and special light lifters. Most of these components are still listed in Ford parts books in Germany, but they are extremely difficult to get in the U.S. (they are sometimes even hard to get in Germany).

One company which does import available RS Capri parts is John Wilhoit's RS West (1360 Gladys Ave., Long Beach, California 90804; catalog $4). John travels to Germany annually to buy parts directly and ship them back. Obviously such components are considerably more expensive than American-made Ford V-6 parts. For example, a complete 2600 RS "tuning kit" costs about $1000. And since these parts are in limited supply even in Germany, waiting time for delivery can sometimes be as long as six months.

Incidentally, Ford of Europe's new Special Vehicle Engineering department recently (1982) introduced a new Capri model in England very similar to the RS 2600 of ten years prior. Using what we would call the Capri-2 body, it is fitted with a German 2800 V-6 with timed fuel injection rated at 160 brake-horsepower. Designated the 2.8 injection, this Capri comes with a front air-dam, rear spoiler, alloy wheels, stiffer suspension, Bilstein shocks, and Recaro seats. In a British Auto Magazine road test it was timed from 0 to 60mph in 8.05 seconds and ran the quarter mile in 15.93 seconds at 89mph, with a reported top speed of 130mph. However, the price for this special Capri in England in 1982 was over $14,000. Any of the components from this new English/German Capri, including the fuel injection unit, should fit '74-79 2800 Mercury Capris, although obtaining them from England would be both difficult and expensive.

Why the Ford/Capri V-6 has not found favor among performance enthusiasts in the U.S. is a bit perplexing, but it's probably a matter of size. It is too small to compete with our plentiful V-8's—even medium-size ones like Ford 289's and Chevy 283's—and it is too big to compete with the popular 4-cylinders like the Pinto. In American racing there just haven't been many classes that cater to engines in the 2.5- to 3-liter range.

Secondly, though the Ford V-6 puts out substantial horsepower for its size, this very "oversquare" engine doesn't produce tons of torque, especially at low rpm like comparable V-8's. For instance, compare the '74 2800 V-6, rated at 105hp at 4600rpm and 140ft-lbs torque at 3200rpm, with the '75 Ford 302 V-8 rated at 129hp at 4000rpm and 213ft-lbs torque at 1800rpm (both engines being available in the same cars). Though the V-8 is not nearly as efficient or highly-tuned an engine, its cubic inches give more accelerating power off the line in a heavy car. American buyers have demanded this type of power for years—but such thinking is finally changing. Hopefully we are ready to appreciate a small displacement, highly efficient, tight revving, road-racing-type of powerplant

Typical of what can be done on a budget, Marc Beck's bright orange "RAD V6" 2800 Capri street coupe features minor body changes, Centerlines, Dobi/Mulholland suspension and a stock 4-speed. Engine changes include an Offy intake with a 390cfm Holley 4-barrel, Mallory dual-point ignition and Hooker headers.

that can sip fuel around town, yet really stretch its legs on the open highway.

The Ford V-6 is such an engine; but, unfortunately, it was ahead of its time. While we had it, we didn't appreciate it. Consequently, hardly anyone in this country used the Ford V-6 in any sort of performance application, and therefore hardly any performance products were developed for it here. Now that the time is right, it has been pulled from the American market, and the prospects of any future hop-up parts from either the factory or the aftermarket are virtually non-existent.

The good news is that the Ford V-6 is built like a thoroughbred to start with. It has a sturdy bottom end, good oil system, smooth flowing heads, and decent induction and ignition systems. The few bolt-on pieces a modifier might need—cams, pistons, intakes, exhaust headers—are available, though limited, in the aftermarket. Although the Ford V-6 will no longer be available in new cars, a good supply of these engines can be found at decent prices in wrecking yards. They make very good engine-swap material, especially for small imports and other 4-cylinder machines such as the several varieties of mini-trucks.

Experts on the Ford V-6 are few and far between. While developing the Dual-Port intake manifolds for this engine, chief engineer Ollie Morris of Offenhauser Equipment spent considerable dyno time with a modified 2800. Turbocharging kits have been designed by Spearco and Ak Miller for these engines. One builder who has considerable experience with the

Ford V-6 builder Jerry Spotts says he has yet to encounter any problem with the stock bottom end, even under severe use.

A 60-degree V-6 slips into a mini truck like it belongs there. Mark Martin used a '73 Capri 2600 with a C-4 automatic to swap some power into his lowered and flared '73 Courier. The engine uses Ford electronic ignition, stock headers, an Isky cam and a non-staged Weber 2-barrel (38 DGAS).

Ford V-6 is Jerry Spotts of Champion Muffler in Whittier, California (affiliated with Ak Miller Enterprises). Jerry developed the Ak Miller turbo system, and has built dozens of these engines in both turbocharged and normally-aspirated versions. Pinto/Capri Performance in Campbell, California, has built many Ford V-6's for street and racing applications. They stock, or can get, most performance parts available for this engine; however, they are not a mail-order business and do not offer a catalog.

A mail-order company which does carry a full line of Capri accessories, including a fair stock of engine components, is Dobi (320 Thor Pl., Brea, California 92621; catalog $1). For the V-6 they offer headers, exhaust systems, ignitions, clutches, camshafts, valve covers, timing gears, oil pumps, induction systems, valves, bearings, and TRW pistons for both 2600's and 2800's.

If you are a Capri owner, a good way to keep in touch with others interested in modifying or racing these cars is to join Far West Capris, a club in California devoted to 4-cylinder and V-6 Capris. You can contact them by writing Jim Razor, 1317 Standish, Arcadia, California 91006.

BLOCK, CRANK, RODS

The performance enthusiast needs to do little, if anything, to the bottomside of the Ford V-6.

Besides the 60-degree angle between cylinder banks, another unusual feature of the Ford V-6 block is extra material on one side of the bottom end, making the crank appear offset to the left. Actually the pan rail and lower block are offset to the right. This extra material accommodated the counterbalance shaft in the V-6's companion V-4, but the shaft is unnecessary in the six. Since both engines were designed to be made from the same tooling, the "bulge" remains in the V-6's.

The 2600 and 2800 blocks are essentially the same, the larger engine having a bore of 3.66 inches (93mm), the smaller a bore of 3.545 inches (90mm). (This German-made engine is all metric, but we will use U.S. equivalent dimensions where applicable.) The easiest way to tell the engines apart is to check the cylinder heads,

The deck area on the Ford is minimal and perforated with water passages. Head sealing can be a problem, especially on turbocharged motors. Note the staggering of cylinder banks for the 6-throw crank in the 60-degree design, plus the very tight valley chamber.

The crankshaft appears offset in the Ford, but actually the block and pan rail are extended on the left side, a carry-over from the German V-4 which has a counterbalance shaft located there. Obviously this adds extra weight.

the 2600 having two exhaust outlets per side, the 2800 three. Although the heads will bolt to either block, a minor difference in water passages prohibits this swap. To identify a bare block, you could match the appropriate head gasket to the head surface.

The two engines also differ in stroke length, the 2600's being 2.630 inches (66.8mm), the 2800's being 2.7 inches (68.6mm). Both engines use the same sturdy six-throw, four main-bearing crank. For the longer stroke, Ford offset-ground the rod journals to a smaller diameter (2.126 inches in the 2600, 2.047 inches in the 2800). The 2800 crank also has a larger diameter front flange, requiring a different front cover on this engine. There appears to be little difference in strength between cranks, neither unit showing weakness. For super high-performance applications the usual crank preparation of straightening, Magnafluxing, Tufftriding, journal polishing, and oil hole chamfering should be considered, but these measures are probably not necessary for most street or mild racing build-ups.

The one weak component of the bottom end is the connecting rod bolts. Since no comparable SPS-type bolts are available to fit these rods, your best insurance is to install new rod bolts every time the engine is torn down, and it would be wise to have these bolts tested for cracks and/or hardness before installation. For high-performance or racing, have the rods

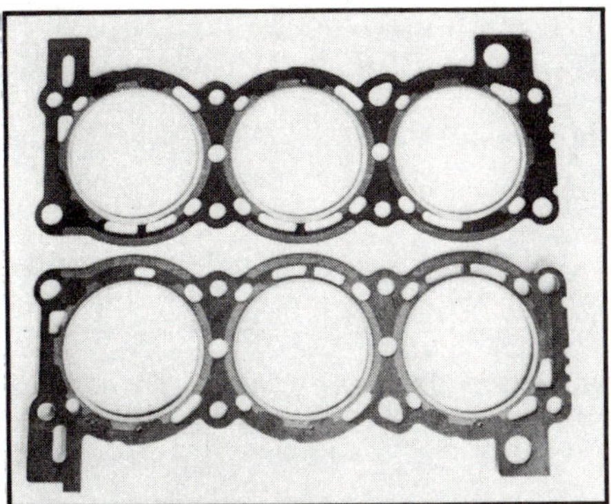

Water passage differences between the 2600 and 2800 are not major, but enough to prevent cylinder head interchange, as shown by this comparison of head gaskets.

polished on the beams, shot-peened, and Magnafluxed. Finally, have the crank and reciprocating assembly balanced in the usual manner.

Most builders have found little problem with the stock Ford V-6 oiling system, including the stock pump. Pinto/Capri Performance feels the engine could use more oil volume, and the TRW high-volume oil pump, part

number 50130, would be a good choice, if you can get it. A high-volume pump is also available from Melling, part number M-87-HV. Engine swappers will be glad to know that Capri oil pans have a rear sump, while Mustang/Pinto pans have a front sump. They're interchangeable, along with the appropriate pickup tube.

Obviously the longer-stroke crank could be swapped into the 2600 block using the 2800 rods. But the minimal displacement increase wouldn't be worth the expense or effort. Swapping the 2800 pistons into the 2600 would require a .115-inch overbore—considerably more than recommended as safe. The factory recommends a conservative .040-inch overbore on the 2600, .020-inch on the 2800; .060-inch should be about the limit on either block. Overboring will be limited somewhat by aftermarket piston sizes available.

PISTONS

Replacement pistons for the 2800 should be easy to obtain in a range of

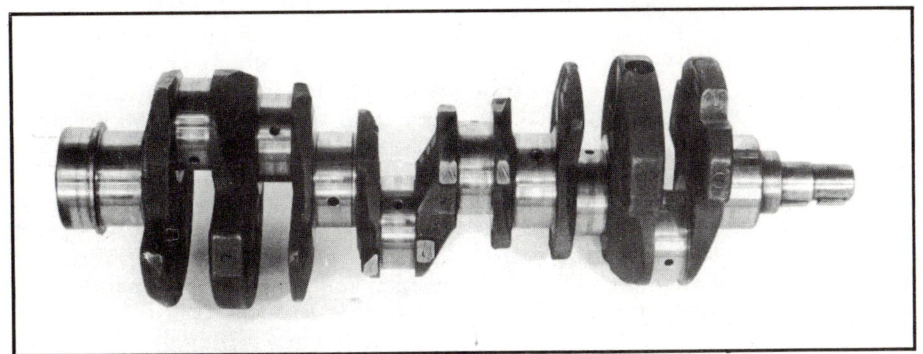

The stock Ford crank is plenty strong and should cause no problems. Give it the usual prep and run it.

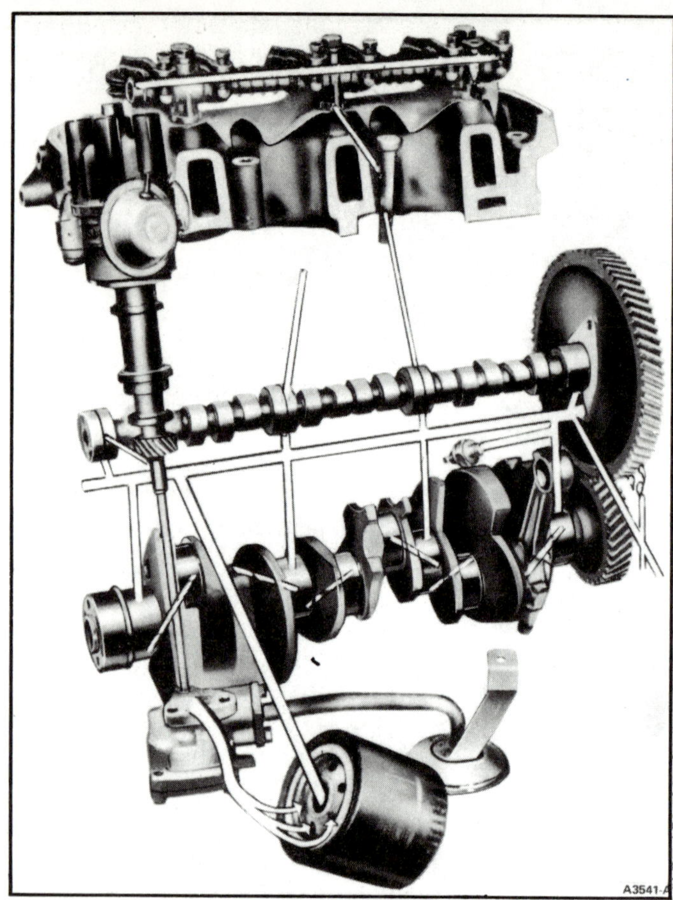

The Ford V-6 oil system features a separate main gallery feeding main and cam bearings; the rockerarms are oiled through the shafts. All engines use mechanical lifters.

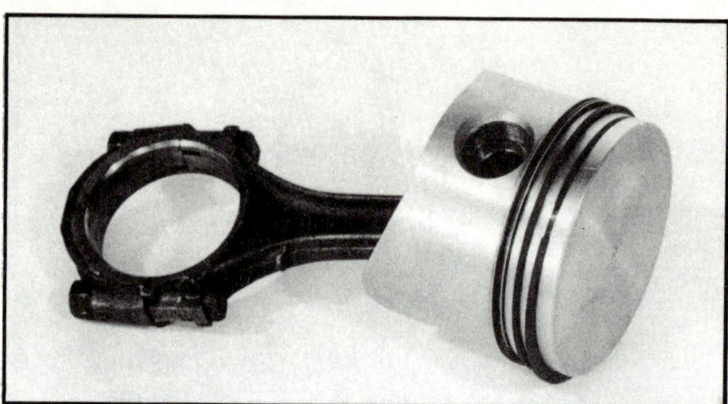

Over-the-counter forged pistons are not available for either Ford V-6, but by boring the 2600 engine .030-inch oversize, you can use forged TRWs for a 2000cc Pinto—after trimming the tops slightly.

Cast TRW replacement pistons for the 2600 (left) and 2800 (right) may be hard to locate, but can be mail-ordered from Dobi.

brands and oversizes. TRW offers a cast aluminum version (part L1270F) in standard and .030-inch oversize. Replacements for the 2600 might be harder to locate, though TRW still lists part L1255F for it. Since pistons are more readily available for the 2000cc Pinto four-cylinder, a trick used by Jerry Spotts is to swap these into the 2600 V-6. The standard bore on the Pinto is .030-inch larger than the V-6, so these pistons will fit in a 2600 bored .030-inch oversize. The Pinto piston also has a slightly taller deck height, and will hit the V-6 head if not trimmed. Cut .030-inch from the top for approximately 9:1 compression; .060-inch for 8:1. These figures may vary slightly depending on piston brand used.

Always be sure to measure both pistons and cylinder bores for proper clearance, and check piston deck height and valve clearance at TDC on any performance engine. To determine exact compression ratios, you'll have to measure chamber, cylinder and piston-crown-volumes and go through the requisite math.

Forged pistons for the Ford V-6's are harder to come by. By swapping Pinto pistons into the 2600, you can use TRW's forged flat-top, part L2395F. Arias pistons (Gardena, California) will make forged pop-up pistons for the 2800 to give a maximum of 10:1 compression. The valves must be recessed approximately .040-inch in their seats for clearance with such pistons. Of course, most of the high-performance piston manufacturers can make custom forged pistons to fit particular applications, but they can be expensive.

HEADS

The major difference between the 2600 and 2800 Ford V-6's lies in cylinder head design. Actually, the combustion chamber and port shape of both heads are basically the same, and both use the same size valves (1.56-inch intake, 1.26-inch exhaust). The difference is the port arrangement in the head, the 2600 having three evenly-spaced intake ports and only two exhaust outlets (one pair of ports is siamesed), while the 2800 has wider-spaced intakes and three individual exhaust outlets.

The heads can be swapped right for left, but they cannot be changed from a 2600 to a 2800. They will physically bolt on, but certain water passages do not line up. These passages could be welded up and re-drilled, but this extra effort does not appear to warrant the expense (it would be easier to get a 2800 block). The 3-port head would seem the better design, both in terms of port area and intake manifold distribution. However, several builders familiar with this engine prefer the 2600 head, saying the bigger ports flow too well, if anything. Other builders have found extra horsepower through extensive porting and intake manifold reworking on the 2800.

For most street or mild performance applications you can leave the ports as is. Polish the combustion chambers and slightly enlarge the valve pockets. On a turbocharged engine, the stock valves appear to work well. For a normally-aspirated engine you could swap in slightly larger (1.65-inch intake, 1.42-inch exhaust) Pinto 2000 valves. They will take a little modification, however. The tips must be cut down approximately 7/32-inch (about to the top split-lock groove), then re-heat-treated (heat the tip to cherry red with a torch and quench in oil). A new groove must then be cut below the remaining two for the keep-

The major difference between 2600 and 2800 heads is port layout. The 2600 features a pair of siamesed exhaust ports, while the 2800 has individual ports.

Although valve sequence remains the same on both heads, the middle intake port has been shifted on the 2800 head, giving unequal port spacing and helping to equalize intake manifold runner lengths.

ers. Valve stem diameter should be the same, but check for clearance. Of course, the valve seats will have to be enlarged accordingly.

One problem area for the Ford V-6 is cylinder head sealing. With only eight head bolts per side, it has a tendency to blow head gaskets, especially when turbocharged. Jerry Spotts highly recommends O-ringing the heads on blown engines, cutting the groove so the O-ring wire sits just above the metal lip on the head gasket. Torque the heads to 80 ft-lb, and *always* re-torque them again after the engine has been run in.

CAM AND VALVETRAIN

Both the 2600 and 2800 came with solid-lifter cams and shaft-mounted adjustable rockerarms. The engines have no oil passages for hydraulic lifters.

Iskenderian Cams (Gardena, California) offers two performance grinds for these motors (both accept the same cam). The F6-4 has 260 degrees of duration, .425-inch lift (the stock cam has only .373-inch lift), and is listed as a "torquer." The F6-66 is called a competition grind at 264 degrees of duration and .448-inch lift. Isky also offers high-performance lifters, retainers, keepers, valve springs, and valve stem seals to fit the V-6.

Norris Cams (Van Nuys, California) makes three billet grinds for the Ford V-6 rated at, respectively: 272 degrees/.444-inch lift; 282 degrees/.451-inch lift; and 300 degrees (int.), 306 degrees (exh.)/.495-inch lift (duration measured at .020-inch lifter rise). The middle profile seems quite popular with Capri slalom and street racers. Norris also offers a special, lightweight, hollow lifter for this engine, as well as valve springs, aluminum retainers, and chrome-moly tubular pushrods. You may find Ford V-6 grinds listed by other cam manufacturers, but they aren't plentiful. Of course, any performance camshaft company can regrind your stock cam to a special profile if you can talk them into doing it.

When installing a higher-lift cam in a Ford V-6, be *sure* to check valve-to-piston clearance at top dead center (.100-inch minimum) with the head torqued down on the block and the cam dialed in. Combustion chamber room is tight on this engine, plus it likes to rev, which can lead to valve float. If more clearance is needed, you can notch the pistons or recess the valves in their seats.

You should always install new lift-

Valve size and chamber shape on both heads is the same, as is the bolt pattern, but differing water passages prohibit head interchange. The 2600 head shown here has been cut for O-rings for a turbocharged application.

Oversize valves are not available for the Ford V-6, but you can modify Pinto 2000 valves (shown on right, compared to stock V-6 valves) to fit the V-6 heads.

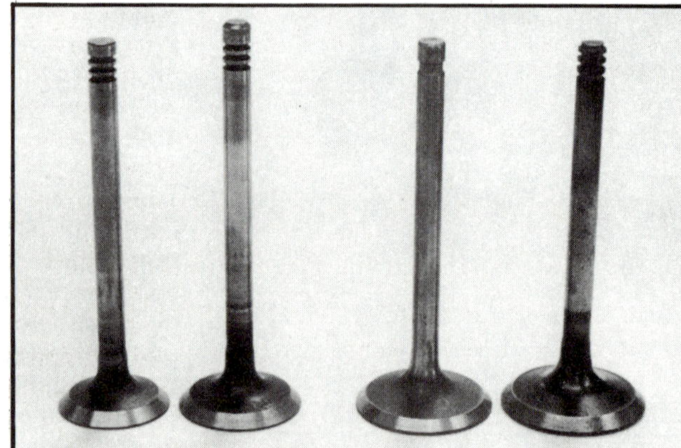

The only custom valve covers made for the Ford V-6 are available from Dobi in Brea, California.

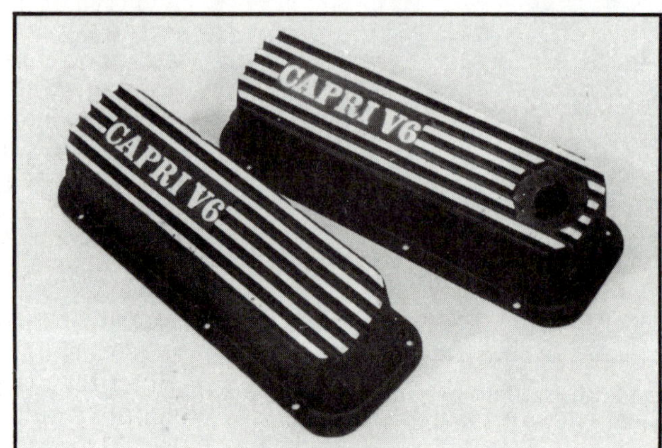

ers with a new cam. However, the tight quarters of the V-6 valley chamber requires removal of the heads before the lifters can be pulled completely out of their bores. This makes a cam change in a Ford V-6 a major operation if the engine isn't already torn down. If you *really* don't want to pull the heads off, you could install the new cam by pulling up all the stock lifters and slipping the cam in place. However, this is a risky procedure (if any of the old lifters are worn, they may ruin the new cam), and we don't recommend it.

Even if you aren't going to install a high-performance cam, we would still recommend using the Isky springs (part 905-D, 1.430-inch o.d., coil spring with damper), or a similar type. The stock springs are too weak, especially for an engine that likes to make horsepower at high rpm. With these springs you might have to drill out the mounting holes in the rockerarm oil-baffle plate and shift it slightly for clearance.

Another weak component of the valvetrain is the stock fiber cam gear. TRW makes a good replacement set with an aluminum cam gear, part TG2760 SA for the 2800, TG2762 SA for the 2600. This gear has an arrow and a couple of dots on it. The dot directly below the keyway is the timing mark; align it to the dot on the crank gear. When the cam is installed in the engine, both the cam and the crank keyways should be pointing toward each other.

INTAKE, MANIFOLDS AND CARBURETORS

The extremely narrow distance between cylinder banks on the 60-degree V-6 leaves little room for intake manifold runners. Designing a smooth-flowing intake was one of the major problems during this engine's development. We would assume the major reason for the head change on the 2800 was to move the center intake ports farther apart in an attempt to equalize runners. Like the other V-6's, this engine is designed (by firing sequence) to use a simple "dual-log" type manifold, the right half feeding the right bank of cylinders, the left half the left bank. A 2-barrel carb could then be mounted transversely atop the manifold with one barrel feeding each "log." This is how the European (and some of the later U.S.) Ford V-6's were set up.

For most U.S. applications, however, the engine uses the Holley/Weber 2-stage 2-barrel with primary and secondary venturis. Since it runs on only one barrel at low and mid range, with the secondary cutting in for acceleration, this carb is mounted "sideways" on the manifold by means of a 90-degree adapter. It looks like an inefficient afterthought, but some builders say it works just fine, if not better than other designs. The Holley/Weber is a good economy carburetor with adequate performance potential when the secondary is actuated. This carburetor should not be mounted without the adapter, since the primary throttle would feed only half the manifold.

If you want to eliminate the sideways adapter, good performance bolt-ons are the Weber models 38 DGAS 3A or 40 DFAV 1, as used on the engines in Europe. On these 2-barrel carbs both throttles operate in unison, unlike a progressive 2-stage. They will bolt directly to the earlier Ford V-6 manifolds, but you will have to fabricate throttle linkage to fit. On cars with automatic transmissions, you may have to add a short aluminum spacer with a vacuum fitting.

The Weber carburetors work excellently when tuned properly for a specific application. For use on the 2600 or 2800, the press-in idle air-corrector bushings (see photo, pg.122) may have to be changed in size. Proper float-level adjustment is also critical, and fuel pressure at the carb must be kept below three pounds maximum (use a fuel regulator). The adjustment of the carburetor should probably be left to someone experienced with Webers. These carburetors are available through any Weber dealer, such as BAP/Geon import parts stores.

For 2800's, a less expensive combination for a large "non-progressive" 2-barrel would be to use the intake manifold from a '76-77 Mercury Capri II V-6, which came with a wide-base

Isky Cams offers two performance grinds for the Ford V-6, along with springs, retainers and lifters.

Motorcraft 2-barrel and will accept a Holley 350cfm model 2300 performance carburetor. This swap has been used successfully by Pinto/Capri Performance on normally-aspirated street-type engines. '79-80 Capris came with a variable-venturi carburetor—a good idea that wasn't quite perfected. They usually suffer from sticking throttle slides after some

On any performance Ford V-6 it is wise to replace the stock fiber timing gear with this TRW aluminum gear set.

miles. It can be swapped directly for the Holley 2300.

As far as the aftermarket goes for intake manifolds, you have one choice—Offenhauser Equipment (Los Angeles, California). They offer their patented Dual-Port intake for both the 2600 and the 2800 in either a progressive 2-barrel or a 4-barrel version. The Dual-Port is actually two manifolds in one, sort of like a sandwich with a primary circuit on the bottom and a secondary circuit on the top. The idea is to keep air/fuel velocity high in the small primary runners, then dump in extra volume when the secondary side cuts in for acceleration or high speed. The 2-barrel version accepts the stock 2-stage Holley/Weber carb, and some builders debate whether it is an improvement over the stock manifold. The nice part is that the Offy manifold is made in two pieces, and the 2-barrel top can be swapped for the 4-barrel top without unbolting the bottom from the engine. This way you can run the 2-barrel for optimum economy, then bolt on the 4-barrel for racing in a matter of minutes.

The Offy 4-barrel top is designed to take a Holley 390cfm carb (part R6299AAA). After extensive dyno and drag tests on a 2800 engine, they suggest the following modifications to this carburetor for this application: change the primary main jets from #48 to #52, the secondary main jets from #50 to #53, the power valve to part 25BF 237A-65, and drill the power valve ports in the primary fuel metering block from .021-inch to .040-inch.

IGNITION

Other than adding one of the many electronic ignition conversion kits available on the market these days

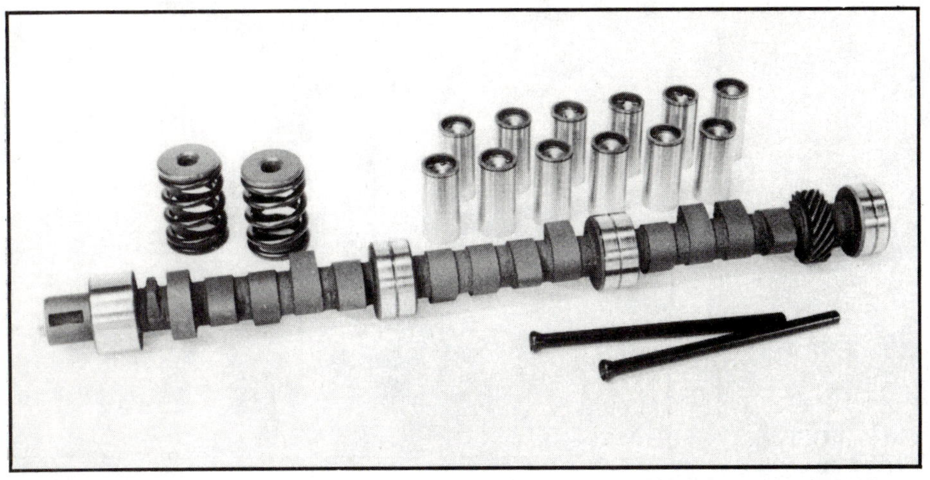

Norris Cams is one of the few companies that specializes in Ford V-6 grinds. They offer three billet profiles, competition springs, tubular pushrods and special lightweight hollow lifters.

Although the earlier Ford manifold appears to have been designed for a standard, transverse-mounted 2-barrel (which it was—a non-progressive carb was used on European models), an adapter to turn the carburetor sideways is necessary to mount the 2-stage Holley/Weber used in the US. This is an excellent performance and economy carb which has one primary bore and a slightly larger secondary that cuts in at full throttle. The adapter, which spaces the carb from the manifold, may actually improve fuel distribution.

(such as MSD, Mobelec or Clifford/Jacobs), you have two choices when it comes to an ignition system for your Ford V-6—the factory points-type distributor or the Ford electronic ignition system used on '76 and later engines. The electronic system, complete with its big blue cap and heavy-duty wires, is probably the better choice. Get the distributor with the proper "box" to match (they vary from year to year). This distributor may have too quick an advance curve, so you may want to install slightly stiffer springs to slow it down a bit. This is a trial-and-error operation, since the compression ratio, cam, and induction system on your particular engine will dictate the proper distributor advance.

The earlier breaker-type distributor has a tendency to cause the breaker points to "float" at high rpm, but it features a dual-diaphragm vacuum advance system, which can be used to advantage on turbocharged engines. A few breakerless units also came with this set-up. The retard (rear) orifice on the canister can be connected directly to the intake manifold, with a small, rubber hose and it will retard the spark when turbo boost cuts in, thus reducing detonation. The Ford electronic ignition components will fit on the earlier distributor, so this might be the best combination for turbo-charged applications. If you are running the dual-diaphragm distributor on a normally-aspirated engine, you'll probably want to disconnect the retard side of the canister unless you experience "pinging" in the engine. Distributors are interchangeable between the 2600 and 2800.

EXHAUST

The short, cast-iron exhaust manifolds that come on the 2600 are pretty efficient for their size, and would work nicely for many engine swaps. The 2800 manifolds vary with engine application, but most are of a smooth-flowing design. The cast-iron manifolds will give decent performance while keeping engine noise to a mini-

Several Weber 2-barrel carburetors can be bolted on or easily adapted to the Ford V-6 intake. The 40 DFAV-1 Weber, shown here, is synchronous (both barrels operate continuously), has 40mm throttle bores and is similar to the production German carburetor for the V-6.

Weber carburetors are fully adjustable, but are unfamiliar to most American mechanics. To tailor off-idle performance (eliminate off-idle stumble), you may have to change the size of the press-in "idle-air corrector bushing" (shown).

The '76-77 Capri II 2800s used a bolt-on carb adapter to mount a large Motorcraft 2-barrel. This manifold will also accept Holley performance 2-barrels.

mum.

Only a few companies make tube headers for the Ford V-6. Hooker makes 2800 headers to fit Capri II's, Pintos and Mustangs. Thunderbird Products (Rancho Dominguez, California) makes one set for 2600's to fit Capris, and ten different sets for 2800's to fit Capris and Mustangs. The 2600 headers are unique in featuring 1.5-inch tubes for the single exhaust ports, 1-5/8-inch tubes for the siamesed ports. The extra long length of these headers should give the little motor a bit more bottom end torque, which it could use. T-Bird's model 6445 header for the 2800, also designed for Capris, would make a good engine-swap exhaust system in a variety of other chassis. Check your local speed shop or parts store for other header possibilities; a few other companies (such as Tri-Mil in Los Angeles, California) have listed Capri or Mustang applications.

CLUTCH AND TRANSMISSION

The weak link in the Capri V-6 power train is the transmission. The German Hummer 4-speed, although it has a nice gear selection, has a fragile cluster that literally "goes away" behind a hopped-up motor with any muscle. Automatic versions of the Capri came with a C-3 automatic from Bordeaux, France, which is a lighter-duty unit than Ford's C-4.

2800's as used in Pinto, Mustangs, and '79 and later Capris came with either the Ford C-4 3-speed automatic or the Borg-Warner SR-4 4-speed stick (automatic only in the Pinto). The C-4 is an excellent trans, especially for this size engine, and it can be

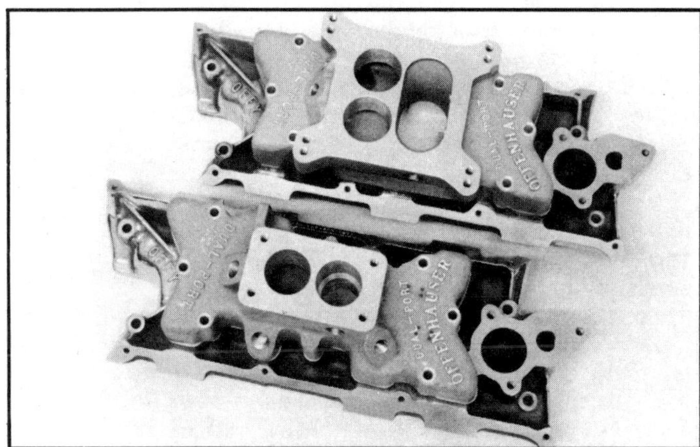

The only aftermarket intakes for the Ford V-6 are these Dual Ports from Offenhauser—available with either a 2-barrel or a 4-barrel top (or both). Photo of bottom halves demonstrates the difference between a 2600 manifold (top) and a 2800 manifold, which will not interchange.

Ford V-6's have used a variety of ignitions including point-type (right), small-body breakerless (middle) and large-body breakerless (left). All will fit both 2600's and 2800's.

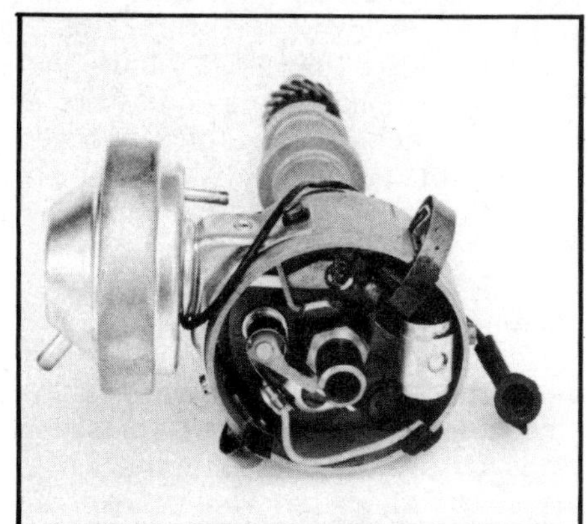

On engines fitted with a turbocharger the Ford dual-diaphragm, vacuum-advance canister—used on breaker-type distributors—can easily be hooked up to retard ignition lead under boost.

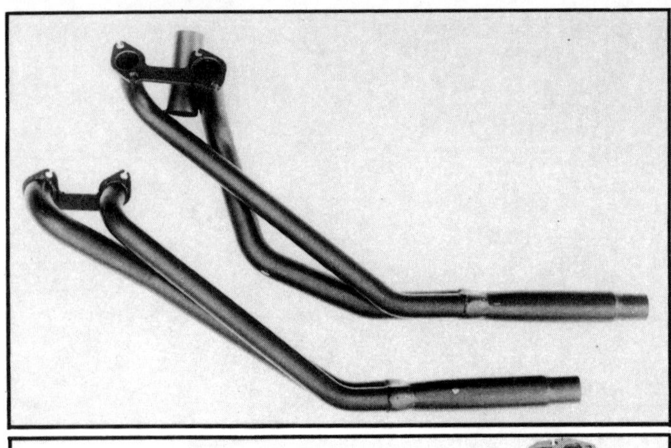

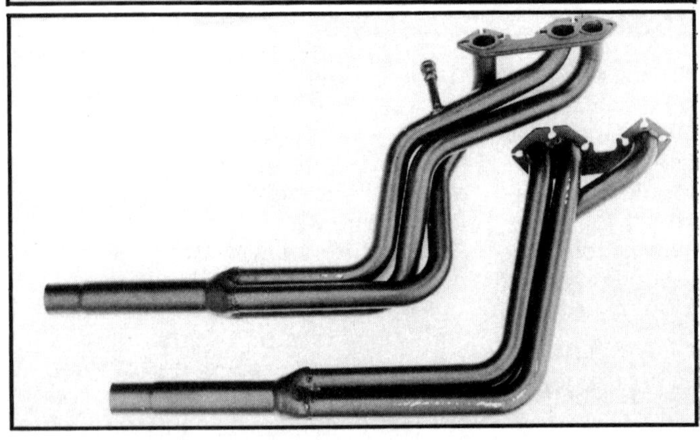

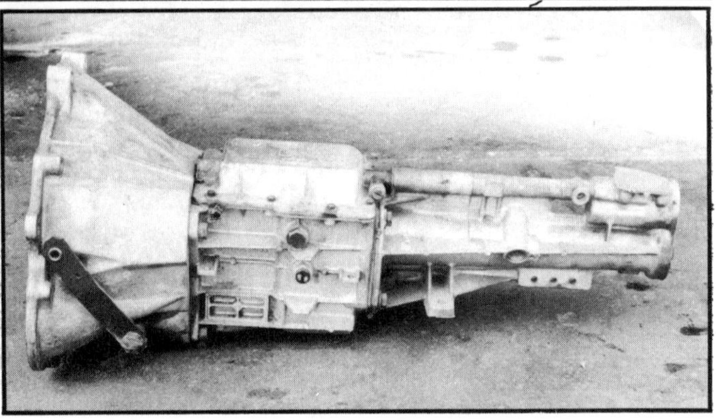

Thunderbird Products, among others, makes several styles of headers to fit Ford V-6s in various body styles. Their 2600 Capri header (top) is unique in having a larger pipe for the siamesed exhaust port.

The German 2600 4-speed (top) is small and light, but it is also fragile. It bolts to the bellhousing from the inside, requiring removal of the bellhousing to remove the trans. The 2800 4-speed (below) is much stronger and has an aluminum case, but the long tailshaft with the gear lever at the end makes it a difficult fit in many engine swap situations.

readily beefed up with kit parts available from B&M, Transgo, and other companies. The SR-4 4-speed is stronger than the Hummer, and is obviously the better choice for a stick shift. However, later versions ('79-up) use a wide-ratio gear set with a 1:1 third and 1:0.8 overdrive fourth. This transmission is obviously great for highway economy, but the giant step between second and third gears leaves the little V-6 gasping for breath. Remember, this engine is designed to work in a relatively narrow high-rpm band.

Both 2600's and 2800's have the same bellhousing flange on the block. To swap transmissions, change bellhousing and all. Since all Capri I and II models (through '77) were completely German-built, they use European (i.e., metric) sizes for transmission, pilot bushing, ring gear, starter, and so on. All Pintos, Mustang II's, and the "3rd generation" Capri use the German engine mated to American transmissions. To further complicate the issue, the 2600 has a shorter crankshaft flange than the 2800 and there are "short" and "tall" flywheels to make up this difference. Plus some versions used the "short" flywheel with a spacer. If you are going to swap trans-

missions, you are going to have to do some mixing and matching of components, and it will probably have to be a trial-and-error process. Be sure that the pilot bushing matches the transmission input shaft diameter; likewise, the spline pattern on the clutch disc must match the transmission input. The starter gear teeth must match those on the ring gear of the flywheel or flex plate, and the spacing of the flywheel must be proper for starter engagement. Finally, remember to use metric bolts in metric-threaded holes. The best way to avoid problems is to get all matching components from the same engine when you are at the

wrecking yard: bellhousing, flywheel, clutch, starter, pilot bushing (buy a new one of these), and transmission.

TURBOCHARGING, SUPERCHARGING, NITROUS-OXIDE

Since few performance products are available for the Ford V-6, many builders have wisely turned to the recently-popular horsepower bolt-ons of turbocharging, supercharging, or nitrous-oxide injection to increase the performance of this engine. Any of the three is a practical way to effectively boost the power of this V-6. With one of

The "tall" V-6 flywheel is shown at right, compared with a "short" flywheel that has been fitted with a spacer, which makes them both the same "height." Some combinations use the short flywheel without the spacer for proper starter engagement.

Ak Miller's turbo kit for the 2600 uses a modified stock exhaust manifold to mount the blower, a cast aluminum, water-heated adapter for the stock carb, and an elbow tube to feed into the stock intake manifold.

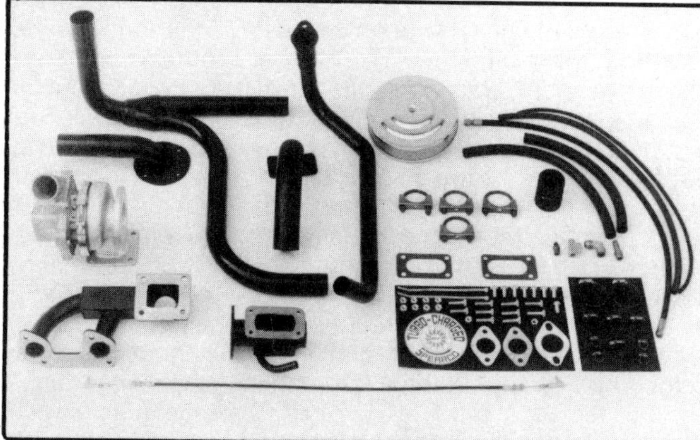

This Spearco kit for a 2600 uses a Rajay turbo and a fabricated exhaust system.

Jack Schorr of Independence, Missouri, bested all cars to win the open division of the Streetkhana at the 1981 NSRA Street Rod Nationals. His homebuilt '23 T roadster uses a '76 Capri II 2800 and C-4 automatic for power. The engine is fitted with a 3-71 GMC blower and BDS drive, O-ringed heads and modified Hooker headers.

Craig Railsback of Blower Drive Service is shown mocking up the prototype for their Magnacharger installation kit for the Ford V-6.

these systems you'll want to add headers (except for turbos), a high-energy ignition, and stiffer valve springs—all of which are available. The stock camshaft seems to work well with a pressure boost, as do the stock valves and ports. The U.S. versions of the Ford V-6 came with 8.2: to 8.7:1 compression, which is just about right for a moderate-boost blower. The stock pistons could live with moderate boost (around seven pounds maximum), but forged pistons are always a good recommendation for a blown or nitrous-assisted engine. O-ringing the heads is also recommended on these engines for such applications.

Ford V-6 turbocharger kits are available from Ak Miller Enterprises (Pico Rivera, California) and Shelby Spearco (Inglewood, California), though the Spearco kits are in jeopardy of being discontinued. Ak Miller uses an AirResearch TO457 turbo for the 2600, a TO468 for the 2800. Blow-through or draw-through kits are available for either engine, using the stock carburetor. The draw-through applications seem to be more popular. However, blow-through set-ups on these engines are relatively simple since either the Holley/Weber or the Motorcraft stock 2-barrel carbs can be fitted with a "bonnet" for turbo pressure, without having to be further sealed or enclosed in a "boost box." Spearco currently lists (as of June '81) kits for 2600 Capris and 2800 Mustang II's using the stock carburetor, and a kit for the 2800 Capri with a side-draft Dellorto dual-throat. Spearco kits are of a draw-through type using Rajay turbos, and produce 7 to 9 pounds boost. These or other companies can also supply necessary hardware to build your own turbo application.

Although turbocharging appears to be the most popular modification to the Ford V-6, its one drawback is that it is most effective at higher rpm. Since this little engine likes to make horsepower in the high rpm ranges, the addition of a turbo makes it a real freeway screamer. A positive displacement blower like a Roots supercharger, however, would help more on the bottom end, where this V-6 could use some beans. An excellent blower of this type, suited to the size of the Ford V-6, is the Magnuson MC-80 Magnacharger (Magnuson Products Development, Santa Ana, California). A kit to adapt this blower to the V-6 is being made in limited quantities by Blower Drive Services (Whittier, California), using the base of an Offenhauser intake manifold with a specially-machined bolt-on top. Using a larger supercharger, such as a GMC 3-71 or 4-71 is more difficult on the 60-degree V-6 because of lack of room between cylinder banks, but a few such installations have been performed.

Nitrous-oxide injection might not be as impressive looking as a turbo or blower, but it can be nearly as effective. Today's systems are safe, dialed-in, and inexpensive compared to the other two. An under-the-carb base plate is the easiest way to add nitrous to the Ford V-6, and recent tests seem to indicate that such systems are even more effective than direct-port injectors plumbed under the manifold (given the same flow capacities). All nitrous companies offer base plate kits to fit Ford V-6's, and most will tailor-make more exotic set-ups. If you are primarily interested in instant power in large quantities for short periods of time (as in racing), nitrous-oxide is probably the most practical way to hop-up the Ford V-6. But it only works when you push the button. If you want to improve the full-time efficiency of the engine for increased power as well as economy, a supercharger or turbocharger would be the better choice.

FORD'S NEW 90-DEGREE V-6

Just as this book was going to press Ford released an all new 90-degree V-6 engine which had been under development for nearly two years. Although this 3.8-Liter (231 cubic inch) motor is initially being installed only in 1982 Lincoln Continentals and Mercury Cougars, where it will receive little attention as a performance powerplant, it appears to have considerable performance potential and should have a significant impact in the V-6 ranks as more of these engines reach the market.

Despite the fact that it is being referred to as the "Essex" V-6 (it has no similarity whatever to the Essex 60-degree V-6 produced for years by Ford of England), this is an entirely new American-made engine. Unlike the Buick and Chevrolet 90-degree motors, which were developed directly from existing V-8 patterns, the Ford 90-degree has a modern, lightweight, cast iron block designed specifically to accomodate the 30-degree split throw even fire crankshaft. Consequently the fore and aft spacing between right and left cylinder banks could be shifted to allow for a wide, strong flange between adjacent rod journals on the crank, and to align cylinder bore centerlines with the rod journals. Thus the new Ford engine does not have to deal with the piston/rod/journal offset problem common to the other 90-degree V-6's. The cast nodular iron crankshaft appears stout, with 2.519-inch main and 2.3107-inch diameter rod journals. The connecting rods are forged steel, with a rod bearing width of .660-inch.

Although the overall Ford design does appear to resemble the even fire Buick, even to the aluminum front cover housing the distributor, fuel pump, and an external oil pump (some Buick builders are already shouting, "copy!"), there are other significant differences. Foremost is weight savings. At 298 pounds complete, the Ford is more than 60 pounds lighter than the Buick. The modern thinwall block is trim, having no extended skirt like the Buick. But the big news is the aluminum head design. Obviously patterned after Ford's highly successful, free-breathing Cleveland V-8 heads, they feature huge oval-shaped intake and exhaust ports with 1.77-inch intake and 1.45-inch exhaust valves. Unlike the Clevelands, the valves are parallel rather than canted, giving a

FORD CAPRI 60° V-6 PRODUCTION SPECIFICATIONS

	2600	2800
Displacement	2.6L (155ci)	2.8L (170.8ci)
Bore	3.545-inches (90.043mm)	3.660-inches (92.961mm)
Stroke	2.630-inches (66.802mm)	2.70-inches (68.58mm)
Years produced (imported to U.S.)	1972-1973	1974-1980
Firing order	1-4-2-5-3-6	1-4-2-5-3-6
Cylinder numbering	1,2,3 right bank 4,5,6 left bank	1,2,3 right bank 4,5,6 left bank
Compression ratio	8.2:1	8.0:1
Main journal diameter	2.243-inches (56.97mm)	2.243-inches (56.97mm)
Rod journal diameter	2.125-inches (53.975mm)	2.047-inches (51.994)
Connecting rod length	Not available	5.140-inches (130.556mm)
Valve stem diameter	0.315-inch (8.001mm)	0.315-inch (8.001mm)
Valve head diameter	Int: 1.562-inch (39.675mm) Exh: 1.261-inch (32.029mm)	Int: 1.562-inch (39.675mm) Exh: 1.261-inch (32.029mm)
Valve lift	.373-inch	.373-inch
Valve clearance (cold)	Int: 0.014-inch (.356mm) Exh: 0.016-inch (.406mm)	Int: 0.014-inch (.356mm) Exh: 0.016-inch (.406mm)
Main bearing clearance	.0005-.0020-inch	.0005-.0020-inch
Rod bearing clearance	.0005-.0020-inch	.0005-.0020-inch
Piston-to-bore clearance	.001-.003-inch	.001-.003-inch
Piston ring gap	.015-.023-inch	.015-.023-inch
	Bolt torque (pound feet)	Bolt torque (pound feet)
Main caps	65-75	65-75
Rod caps	22-26	22-26
Cylinder head	65-80	65-80
Rockershafts	43-49	43-49
Intake manifold	15-18	15-18
Spark plugs	15-22	15-22
Camshaft gear	32-36	32-36
Crankshaft gear	32-36	32-36
Flywheel	45-50	45-50

typical wedge combustion chamber, but the valvetrain remains the same (stamped steel 1.73:1 ratio rockers on individual pedestals). The aluminum pistons are very slightly dished and the compression ratio on 1982 engines is 8.8 to 1.

The aluminum, single plane intake manifold is die cast in two pieces and then welded together. Apparently the engine will be produced with either a variable venturi or a standard Motorcraft 2-barrel carburetor depending on application. California models will be fitted with a Microprocessor Control Unit (MCU) to regulate air/fuel mixture and a knock sensor to retard ignition timing, but "49 state" models will not be computer controlled. All engines will be fitted with Ford's Duraspark II breakerless ignition system.

Since this new V-6 had barely hit the showrooms when this was written, we have nothing to report in the way of performance applications or testing. Since the motor is currently available only in luxury models, and since the attention of Ford's reborn Specialty Vehicle Operations (that is, performance) department has been directed to the four cylinders and the 302 V-8, we expect little in the way of performance development for the "Essex" V-6 in the near future. And until the motor is made available in more popular or performance-oriented cars, we doubt that the aftermarket will tool up for specialty parts for it. However, this new V-6 does appear to have considerable potential as a strong, light, compact performance powerplant, even in stock form or with a few home-grown modifications such as a carburetor change, reground camshaft, custom-built headers, ignition re-curve, minor headwork, or even the adaptation of a turbocharger, supercharger, or nitrous oxide injection.

The new 90-degree Ford 3.8-liter V-6 looks a lot like the Buick; it even has a front-mounted external oil pump. The front cover, intake manifold and heads are aluminum. The valve covers are plastic.

Since the new 90-degree block was designed expressly for the V-6, the cylinder banks could be offset to provide correct alignment of rods and pistons on the split-throw crank.

Although the valves are not canted, rockerarm design is the same as on Ford Cleveland V-8s. Aluminum intake accepts a Motorcraft 2-barrel or variable-venturi carb.

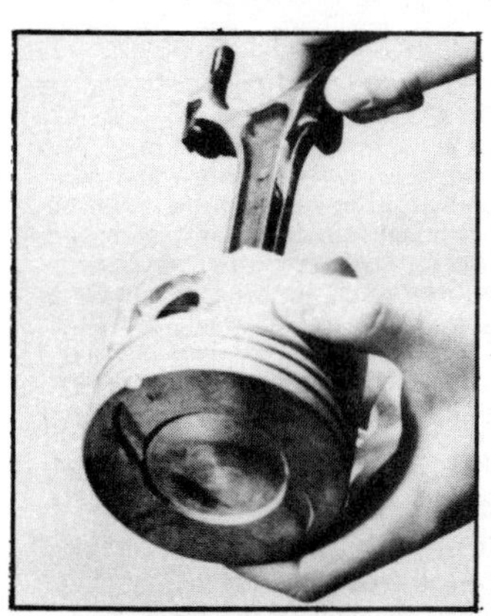

The crankshaft has huge journals and a wide flange between the 30-degree split throws. The block is "skirtless." Pistons are flat tops rated at 8.8:1. Intake and exhaust ports in aluminum head are generous.

FORD 3.8L 90° V-6 PRODUCTION SPECIFICATIONS

Displacement	3.8L (231ci)
Bore	3.80-inches (96.52mm)
Stroke	3.40-inches (86.36mm)
Firing order	1-4-2-5-3-6
Cylinder numbering	1,2,3 left bank 4,5,6 right bank
Compression ratio	8.8:1
Combustion chamber volume (head)	64-67cc
Main journal diameter	2.519-inches (63.98mm)
Rod journal diameter	2.3107-inches (58.692mm)
Rod journal offset (split)	30 degrees
Connecting rod lenght	5.912-inches (150.17mm)
Valve head diameter	Int: 1.77-inches (44.69mm) Exh: 1.45-inches (36.83mm)
Camshaft Duration	Int: 249 degrees Exh: 263 degrees
Valve lift	Int: .415-inch (10.54mm) Exh: .417-inch (10.59mm)
Rockerarm ratio	1.73:1
Valve clearance	Zero
Main Bearing clearance	.0005-.0023-inch
Rod bearing clearance	.0009-.0027-inch
Piston-to-bore clearance	.0014-.0022
Ring end gap	.010-.020-inch